Health and Safety
in the Chemical Laboratory

Special Publication No. 51

Health and Safety in the Chemical Laboratory— Where do we go from here?

The Proceedings of a Symposium Organised by the Professional Affairs Board of The Royal Society of Chemistry, under the Aegis of the Working Party on Professional Affairs of the Federation of European Chemical Societies (FECS Event No. 50), as part of the 1983 Annual Chemical Congress of The Royal Society of Chemistry

University of Lancaster, 12th—13th April, 1983

The Royal Society of Chemistry
Burlington House, London W1V 0BN

British Library Cataloguing in Publication Data

Health and safety in the chemical laboratory
  Where do we go from here?—(Special
  publication, ISSN 0260-6291; no. 51)
  1. Chemical laboratories—Safety measures
  I. Royal Society of Chemistry. *Professional
  Affairs Board* II. Federation of European
  Chemical Societies. *Working Party on
  Professional Affairs* III. Royal Society of
  Chemistry. *Annual Chemical Congress (1983 :
  University of Lancaster)* IV. Series
  363.1'79   QD51

ISBN 0-85186-945-9

Printed in Great Britain by
Whitstable Litho Ltd., Whitstable, Kent

Introduction

This International Symposium (the proceedings of which are published here) was organised by the Professional Affairs Board of the Royal Society of Chemistry. It brought together over 100 participants at the University of Lancaster (United Kingdom) on 12 and 13 April 1983.

The Symposium was the third in a series covering safety conducted under the aegis of the Federation of European Chemical Societies' Working Party on Professional Affairs. The first took place in 1980 in München (Munich) and was entitled 'Safety in Clinical Chemistry Laboratories'; the second 1981 in Tübingen (Federal Republic of Germany) was entitled 'Safety in the University and Teaching Laboratory'.

The Royal Society of Chemistry, a learned and professional body, has its origins in the Chemical Society (formed in 1841) and the Royal Institute of Chemistry (formed in 1877). It is the sole professional qualifying body for chemists in the United Kingdom and has a total membership of around 40,000 individual chemists of which some 10,000 live outside the United Kingdom.

The Federation of European Chemical Societies with member societies in eastern and western countries in Europe is a voluntary association. Its object is to promote co-operation in the field of chemistry in Europe between societies whose memberships consist largely of individual qualified chemists. The Federation seeks to achieve its objects by several means including

- providing a forum for the exchange of opinion on matters affecting chemistry and chemists

- encouraging conferences for the discussion of chemical science and its application

- establishing working parties to survey and report on specific aspects or areas of chemistry in a European context.

The Federation's Working Party on Professional Affairs which meets twice a year is drawn from member societies in 16 countries. It maintains an overview of matters affecting the profession of chemistry in Europe. It produces proceedings of symposia, collects statistics and disseminates information.

Contents

Accident and Dangerous Occurrence Statistics in the United Kingdom

By Mr S. G. Luxon
HEALTH AND SAFETY CONSULTANT (FORMERLY HM DEPUTY CHIEF INSPECTOR,
HEALTH AND SAFETY EXECUTIVE),UK

As our Chairman has indicated, the operative words for this
symposium on Health and Safety in the Chemical Laboratory are
'where do we go from here'.

Perhaps we can better judge in which direction we should go if
we examine in some detail what has been achieved in the past.
Unfortunately much of the available information is of a general
nature because until very recently health and safety legislation,
and hence notification procedures, have not been applicable to
chemical laboratories. The reason for this is primarily historical.

Early legislation in the last century concentrated first on the
exploitation of women and young persons and then on mechanical
safety in the major areas of industrial expansion at that time, i.e.
textiles and engineering. The legislation was accordingly designed
to cover these particular areas of employment. Subsequent
legislation followed this same pattern until by the middle of this
century coverage had been extended to all persons employed in
'Manufacturing industry'. Unfortunately the type of premises as
defined included only certain chemical laboratories, operated solely
for plant-control purposes. The great majority of chemical
laboratories thus remained outside the scope of safety and health
legislation in the UK until the passing of the Health and Safety at
Work etc. Act (HSWA) in 1974. Even then this Act provided only
enabling powers and contained no specific requirement as to
reporting of accidents. It was not until regulations made under the
HSWA came into force in 1981 that any specific requirement to notify
accidents and dangerous occurrences in all laboratories came into
effect.

You will by now have realised that this sad story of neglect means that sound statistics on accidents and dangerous occurrences in chemical laboratories are only now becoming available. Therefore, we must look at more general statistics to see what lessons can be learnt and what directions future action should take. Before doing this however we must have some knowledge of the way in which these statistics have been compiled so as to gain a better appreciation of their significance.

There are two broad areas where reporting has been required over the years: accidents to employed persons and incidents involving dangerous occurrences.

Before 1975 all industrial accidents and dangerous occurrences were reported directly to the Factory Inspectorate. As I have already indicated, after the coming into force of the HSWA, which extended cover to all persons at work and even to the general public, it was accepted that new regulation would be required to widen the scope of the notification procedures.

At the same time, in order to avoid dual reporting to the Department of Health and Social Security (DHSS) and the Health and Safety Executive (HSE), and thus reduce the administrative burden on industry, accidents were divided into two groups -

    (i)   Fatal and serious accidents involving
            fractures or hospitalisation for
            24 hours or more
    (ii)   Accidents involving an absence of more
            than 3 days.

The first group would, because of the urgency of the investigation, continue to be reported directly to HSE. The second, and by far the larger group of more minor accidents would be reported only to the DHSS, who would in turn pass on the relevant information to HSE. As such a system would be based on claims for financial benefit it was believed that non-reporting would become negligible, and hence the statistics more reliable.

Table 1   <u>Reporting of accidents to the central government</u>

        1901 - 1980   Under Factories Acts <u>etc</u>.
                           Over 3 days absence from normal work.
                           'Industrial' Occupations only.

        1981 - 1983   Under DHSS Legislation.
                           Over 3 days absence.
                           All work activities.

        1983 -   New Social Security and Housing Bill
                           Does not require reporting.
                           Consultative Document.

Unfortunately this revised system had been in operation for just over a year when it was overtaken by a more general exercise. In 1980 with the rising administration costs of social benefits it was decided that to streamline the payment system for absences of less than 8 weeks the employer would make the benefit payments. As from 6th April 1983 an accident involving an absence of less than 8 weeks is not reported to the DHSS. Such accidents therefore are no longer notified to the HSE. The net result is that the majority of accidents are no longer reported.

In order to regularise the situation new regulations are in the course of preparation. They will in general attempt to restore the position to that pertaining in 1980 so that continuity of information is assured. It is understood that a Consultative Document will be issued shortly. This will provide a further opportunity to review and refine the system so as to ensure that it meets the general criteria given in Table 2.

Table 2   <u>Criteria for accident statistics</u>

        Unambiguous definition.
        Provide accurate information.
        Comparability - nationally and internationally.
        Sufficient data locally and nationally.
        Assess effectiveness of Safety policy.
        Simplicity of notification procedures.

The health, safety and environment committee of the Royal Society of Chemistry will of course be considering and commenting on this document.

We turn now to the notification of dangerous occurrences. The situation in regard to reporting is somewhat simpler in that no

changes have been proposed, i.e. they continue to be reported
directly to HSE. The definition of a dangerous occurrence has,
however, been extended. Before 1981 only certain very limited
classes of occurrence had to be reported. These were, principally,
serious fires and explosions, failures of lighting apparatus, and
electrical fires. Notification of such occurrences are useful in
that they provide information on serious malfunctions of equipment
or systems of work, which have a high potential for injury but which
occur only rarely.

In order to make further use of the concept the 1980 regulation
carried the notification much further by including fourteen new
classes of dangerous occurrences. Of these three are of particular
interest to us today. In 1981 there were 307 dangerous occurrences
reported in the chemical industry, of which no less than 69% were in
these three classes.

The first and largest class (44%) was Class 8 - 'The uncontrolled
release or escape of any substance in circumstances which ... might
be liable to cause damage to health or major injury to any person'.
This is a wide ranging requirement couched in very general terms
which relies on a large measure of judgement. The key words are
underlined release and liable to cause injury. As with similar
legislative requirements much will depend on the circumstances of each
case, but careful consideration must be given as to whether or not
such an escape could, given customary work patterns, have caused
injury whether or not that injury did in fact occur.

Class 9, which accounted for 11% of all notifications in 1981,
is also new and of importance. It requires the reporting of any
incident in which any person is affected by the inhalation or
absorption of any substance or by lack of oxygen to such an extent
as to cause acute ill health requiring medical treatment. This again
is a wide ranging requirement as medical treatment includes treatment
by the firm's own medical staff.

However, the requirement is more clear-cut and is solely
dependent on whether or not medical treatment was necessary to
alleviate the symptoms.

Class 4 continues the earlier requirement to report any process explosion or fire resulting in a stoppage or suspension of work for more than 24 hours. This accounted for 14% of the reported incidents in 1981.

One other class is of interest, namely Class 10. This overlaps to some extent with Class 8, which we have already considered. It requires any case of acute ill health resulting from occupational exposure to pathogens or infected material to be reported.

These new requirements will provide much useful statistical information of particular relevance to the chemical laboratory over the next 10 years. Having considered the statistical base for the reporting of accidents and dangerous occurrences let us now look at some relevant statistics gathered over the past decade.

The frequency rate* is considered to be a better indicator of performance as it takes account of the hours worked. Table 3 looks at the general trend in respect of accidents over the past decade. It shows a reduction of some 30% over the period.

Table 3   Accidents, all industries

Period	Average number of accidents per year
1972 - 1975	347,000
1976 - 1979	320,000
1980	254,000

A similar trend has occurred in fatal accidents (Table 4), the statistics for which are more reliable because they are all reported. However, since 1979 the downturn in the economy may have affected figures as much as improved accident prevention techniques. The situation in the chemical industry is compared with that in other sectors in Table 5. This shows that incidence rates for all reported accidents in the chemical industry are about the average for all manufacturing industry. It is interesting to note that for agriculture, railways, coalmining and quarrying the figures are some 4, 5, 7, and 10 times as high respectively.

$$* \text{ Frequency rate } = \frac{\text{Total accidents x 100,000}}{\text{Total hours worked}}$$

Table 4   Fatal accidents, all industries

Period	Average number of fatal accidents per year
1972 - 1975	230
1976 - 1979	165
1980	124

Table 5   All reported accidents

Incidence rates per 100,000 at risk

	Yearly averages		
	1976-1979	1980	(Fatal)
Chemicals	4000	3420	(2.4)
Food and drink	4600	4120	(2.8)
Metal manufacturing	6300	4710	(7.2)
Engineering	3800	3050	(3.5)
Vehicle manufacturing	3200	2760	(2.2)
All industry	3500	2860	(2.7)

Within the chemical industry differences between various product groups are given in Table 6. Chemical laboratories are probably most closely related, in general, to Drugs and Fine Chemicals. and are therefore probably at the lower end of the scale.

Table 6   Chemical industry product groups frequency rates 1981*

Heavy chemicals	1.35
Drugs and fine chemicals	0.97
Plastics	1.20
Fertilisers	0.98

*Most recent figures available

Table 7 shows the effect of the size of the undertaking on the accident performance.

Table 7   Effect of size of undertaking

Chemical industry 1981*

Number of employees	Frequency rate	
	Heavy chemicals	Other
Fewer than 100	1.64	1.47
101 - 500	1.54	1.59
501 - 1000	1.41	1.16
Over 1000	0.92	0.87

*Most recent figures available

With increasing size of organisation there is a lower accident
rate in spite of what is probably a higher reporting standard.
This may be due to better management, training, and worker
participation. Although these statistics show general trends,
causation is of more importance than accident prevention. Table 8
shows the causes of accidents in all industries.

Table 8   Causations in all industries in 1981*

Machinery	26%
Transport (Striking Against)	6%
Containers (Muscular)	11%
Building and work surfaces (Falls)	14%
Working environment (Falls)	2%
Fires and explosions	0.5%
Chemicals	2%
Materials and articles (Lifting)	24%
Other	14.5%

*Most recent figures available

It will be seen that 62% of all accidents involve working
practices which are dependent on operator awareness and in respect
of which physical precautions can play only a small part.

When we look at the very latest figures (not yet published)
for 1981 in the research and development area (Table 9) we see that
the same broad causation classes cause the majority of accidents.

Table 9   Non-chemically related accidents in research and
development services 1981*

Type	Number	Percentage of total
Over exertion	131	25%
Falls		
(a) on the level	112)	
(b) from heights	50)	31%
Struck by object	89	17%
Striking object	50	9%
Mechanical	57	11%
All (chemical and		
non-chemical)	530	(90%)

*Most recent figures available

Chemically related accidents account for only 5.6% of the total
(Table 10). To get an idea of what these chemically related
accidents are we must turn to earlier more general statistics in the

absence as yet of a further breakdown of the 1981 figures. These
fall into three groups - gassing accidents, notifiable diseases,
and industrial diseases.

Table 10   Chemically related accidents in research and
           development services 1981*

Type	Number	Percentage of total
Harmful substances	14	2.6%
Extremes of temperature	16	3.0%
Electrical	0	0
Asphyxiation	0	0
All (chemical and non chemical)	530	(5.6%)

*Most recent figures available

Table 11 shows the average numbers of gassing accidents involving
chemicals other than solvents which have been notified over recent
years. Table 12 gives the same information for accidents involving
gassing with solvents.

Table 11   Gassing accidents (chemicals other than solvents)

Chemical	Average number of accidents notified each year
Chlorine	43
Carbon monoxide	16
Ammonia	10
Sulphur oxides	15
Hydrogen sulphide	9
Hydrogen chloride	7
Nitrous fumes	5
Phosgene	4

Table 12   Gassing accidents (solvents)

Solvent	Average number of accidents notified each year
Trichloroethylene	10
Dichloroethylene	5
Formaldehyde	5
Other chlorinated hydrocarbons	6

Table 13 shows chemicals causing notifiable diseases under the
Factories Act, 1961.

Table 13   Chemicals causing notifiable diseases under Factories Act, 1961

	1976	1977	1978	1979	1980
Aniline	35	25	12	13	6
Cadmium	7	1	–	–	–
Chrome ulceration	65	120	65	36	39
Epitheliomatus ulceration	7	15	12	4	5
Lead	31	12	14	8	11
Other	3	1	7	8	3

Table 14 shows diseases qualifying for DHSS benefit resulting from exposure to chemicals.

Table 14   Industrial diseases qualifying for benefit

	Yearly average 1976 - 1980	1980
Lead	31	21
Benzene	3	Nil
Cadmium	6	7
Nitro/amino/chloro- aromatics	2	Nil
Mercury	2	3
Phosphorus	3	1
Other	6	2

So much for the statistics; what inferences can we deduce bearing in mind the need to make the most cost-effective decisions?

1. More can be gained by education and co-operation of the entire work force than by any other factor.

2. Falls, striking objects, and muscular injury account for more than half of all accidents and these can only be reduced by work practice training, better supervision, and an awareness of the danger by all concerned.

3. 'Chemical accidents' account for only a small part of the total but their severity may be greater.

4. Many long-term effects of chemicals are probably not recognised and hence are not included in the statistics. Where there is a delay between the exposure and the onset of symptoms this is even more likely to be the case.

5. The statistics we have been considering are national in
character. In consequence they cannot provide the detail on which
local action can be based and monitored.

Table 15 shows how more detailed statistics compiled by a
single laboratory group can provide additional information. The
first three columns show all accidents for each of three
approximately equal occupational categories while the last column
shows the total of all 3 day accidents. The lower numbers in the
more skilled group indicate the important part that training can play
in accident prevention.

Table 15   Breakdown of accidents reported from January 1973
             to December 1982

Year	Academic staff	Technical staff	Domestic staff	More than three days absence
1973	3	32	51	3
1974	2	14	50	7
1975	4	21	54	2
1976	2	30	39	8
1977	2	32	40	12
1978	7	40	45	7
1979	5	55	58	8
1980	9	44	97	8
1981	5	33	68	10
1982	5	27	69	9
Total	44	328	571	74

Finally, to return to our main theme - where do we go from here?

It is clear that we need more detailed statistics on accidents
in chemical laboratories. It is doubtful whether this could ever
be part of a national exercise because there is no particular reason
to single out this group nationally and a more general exercise
would be too complex.

There is clearly a need for voluntary effort either by
Societies or by laboratory groups to compile more detailed statistics.
Such information would enable more reliable and cost-effective safety
policies to be developed.

In respect of safety all the available statistics demonstrate that training coupled with clearly understood systems of work and a personal involvement with safety on the part of all concerned will achieve more pound for pound than money spent on physical precautions. That is not to say that the latter should be totally neglected, rather that the effort to push forward in the man-management area should have priority.

In respect of health risks there is even less available information and it will clearly be much more difficult to obtain. Although training has an important part to play, identification of the hazard and its containment must be a prime objective.

Perhaps one of the most important aspects of the statistics is their use in identifying hazards and ranking their relative importance.

The next paper shows what can be done in this direction by voluntary effort.

Discussion

Q1.   Does Mr Luxon agree that we should encourage the reporting of 'dangerous occurrences' (as opposed to accidents which are much easier to deal with)? There seem to have been moves in this direction recently.

Luxon -   Yes, I agree. I believe that we are learning lessons in this area.

Q2.   Would Mr Luxon care to elaborate on the lower accident frequencies in large organisations as opposed to small ones?

Luxon –   I had hoped to refer further to this in my
           presentation but time did not allow me to.
           There is considerable debate over this fact
           but there does not appear to be a real answer
           as yet.

           It has been postulated that larger organisations
           tend to have better systems of organisation of
           safety policy, work programmes etc. It is
           interesting that we might expect large organisations
           apparently to have higher rates since they might
           tend to report almost 100% of accidents, whereas
           some very small firms may not report any but the
           most serious accidents. Therefore it is likely
           that the true discrepancy between the rates for
           small and large organisations is actually greater
           than shown by the figures.

           The lower rates for large organisations are perhaps
           a good indication of what can be achieved by good
           management procedures. A further consideration is
           that large organisations tend to have more money
           per employee to spend and can take a longer-term
           view of things. A large firm can spread the
           costs of expensive safety equipment over several
           years.

           These are some of the reasons that have been
           suggested but I would welcome any further ideas
           that people may have.

           It is, of course, one of the few statistics
           that provide us with a lead.

Q3.   Many accidents seem to fall into the category
           of the 'one man trap'. Once the trap has been
           sprung it may affect 1 person or 1000. It
           still constitutes one incident.

Luxon — Thank you for making that point. It has to be said that in a sense we are clutching at straws since there are not really any statistics available, especially for laboratories.

Q4. My experience in teaching students supports the statistics. I find that students from larger firms have usually been well trained in safety consciousness. Those from small firms often show little interest in safety matters. Why is this so?

Luxon — It probably comes back to finance. Large firms can treat precautions on a more impersonal basis whereas small firms tend to see expensive safety measures as immediate on-costs and are therefore less inclined to take such measures.

Q5. What is the role of the insurance companies in the UK in this area? Do higher risk organisations pay higher premiums or do the insurance companies take a more homogeneous view?

Luxon — The situation in the UK is very different from that elsewhere and especially in the USA. Insurance Companies in the UK tend to take a more homogeneous view with only a little attention being paid to a bad history of compensatory payments.

I myself would like to see insurance companies play a bigger part - it would be a good incentive as it is in the USA.

Q6. What can we do to obtain statistics on the long-term effects of chemicals?

<u>Luxon</u> –   This is the intent of the latest UK regulations
which cover occurrences where a person is exposed
to a chemical such that there is liable to be an
injury to health. The words 'liable to be'
suggest a continuing requirement.

This is however a difficult area. Many chemicals
are used and there may be cumulative effects,
synergistic effects <u>etc</u>. which may complicate
matters. I think the solution is to perform
studies on a very wide basis – studies on small
groups pose considerable problems.

Barry Henman will be telling you about the Royal
Society of Chemistry's studies in this field.

Morbidity and Mortality Studies

By Mr B. A. Henman

SECRETARY (PROFESSIONAL AFFAIRS), THE ROYAL SOCIETY OF CHEMISTRY, UK

John Maynard Keynes said 'In the long run we're all dead';
Jonathon Miller, playwright and doctor, said 'Mortality is not
simply an avoidable accident but a natural appointment from
which there is no hope of escape'; Tom Stoppard, playwright,
said 'Life is a gamble at terrible odds; if it were a bet you
wouldn't take it'.

The Shorter Oxford English Dictionary (OED) gives as a
definition of morbid 'of the nature of or indicative of disease'
and of morbidity (in its medical sense) 'prevalence of disease;
the sick rate in a district'. It defines mortality as 'the
condition of being mortal or subject to death; mortal nature
or existence' and as 'the number of deaths in a given area or
period, from a particular disease etc.; death rate'.

Morbidity and mortality studies are very much about disease
rates and death rates; and the practical applications are about
excessive prevalence. The study of the distribution and
determinants of disease frequency in man is known as epidemiology;
the associated term epidemic although popularly used to describe
an acute outbreak of infectious disease is perhaps best defined
in the measured words of the shorter OED "Of a disease:
'prevalent among a people or a community at a special time, and
produced by some special causes not generally present in the
affected locality'." It is said that the United States is in
the grip of epidemics of coronary atherosclerosis and of lung
cancer.

The incidence of both of these diseases has built up to
epidemic proportions over a period of decades. In this context
it should be noted that the proposition that epidemics have to
come about within a period of days or weeks to have that name

(e.g. smallpox in the past or influenza at the present) is no longer considered essential to the definition of epidemics.

What are morbidity and mortality studies and how are they carried out? Such studies can be

- descriptive and concerned with observation of
  distribution of diseases and death

- analytical and concerned with investigation of
  hypotheses suggested by the above

- experimental concerned with measuring the effect
  on a population of intervention in some way or
  another.

This paper is concerned with the descriptive and analytical approaches cited above and will refer specifically to studies being conducted by the Royal Society of Chemistry (RSC).

That such studies can be carried out at all is due to a particularly open piece of legislation passed in 1836 in the reign of William IV. The Birth's and Death's Registration Act, passed in the main to make provision for the registration of births and deaths of dissenters (those who did not wish these events to be registered through the Church), also made it possible for these records to be freely available; and this facility has remained ever since.

It is not all that common for statistics of death to be collected and stored as they are in the United Kingdom (UK). Very briefly the procedure adopted is as follows:

(i)   after the death of an individual a medical
       certificate giving the cause of death is issued
       and signed by a qualified medical practitioner
       together with other relevant medical details;

(ii)   a draft entry of death registration is produced
       on a form by a registrar of births and marriages
       and deaths (an official provided for by legislation)

who transcribes the details provided in the medical
certificate and adds further details provided by
an informant (usually a near relative of the
decedent) regarding the name, sex, marital status,
maiden name (in the case of a married woman), date
and place of birth, final occupation, and usual
address of the decedent together with the date and
place of death;

(iii)   a copy of the form is sent to the death registry
[in England and Wales, the Office of Population
Censuses and Surveys (OPCS); in Northern Ireland
and in Scotland, the respective General Register
Office] and the cause of death coded to the International
Classification of Diseases, Injuries and Causes of
Death (ICD);

(iv)   the form is then stored physically (so that
applicants can obtain a certified copy) and details
are computerized (so that statistical tabulations
can be produced).

Thus it is possible on most occasions to obtain information about
an individual and the cause of death of that individual - all on
one form.

     Other countries have systems which by accident or by design
can tend to thwart the collection of such information. Countries
with a federal system of government with several states in the
federation (e.g. Australia, USA) may collect statistics of death
on a state-by-state basis and may not have centralised registration.
Thus out-of-state deaths may be difficult to trace. Countries may
have laws on confidentiality of personal information which make it
very difficult for statistics of death (other than of a very
general nature) to be obtained. Countries may have systems where
the administrative procedure is such that two quite separate
documents are produced - one a certificate simply to the effect
that a named individual is dead and the other to the effect that
an unnamed individual has died of a certain disease.

Collection of mortality statistics over many years by a country makes it possible for retrospective (i.e. backward looking) mortality studies to be conducted. The degree of retrospection is dictated by the amount of precision that can be applied to the process of searching that leads to the certificate of death applicable to the individual in question being obtained. As exemplification, it may well be helpful here to digress to the mechanics of obtaining certificates of death or details of the current status of an individual from our register offices in the UK - specifically from the OPCS (the registry for England and Wales).

Organisations such as the RSC have very good records relating to individuals who are currently (or who may have ceased to be) in membership. The date of birth and full name are known from application forms for membership; the home address for correspondence is usually known (but in the case of those who have resigned from membership it may not be up-to-date); a date of death of those who died in membership is usually known, through correspondence with a near relative. As regards the date of death there is little chance that it will be known for ex-members - or even that they have died.

Most people in this country are registered with a general medical practitioner and because of this have a National Health Service (NHS) number. General medical practitioners in the NHS are paid on a capitation basis, i.e. on the number of patients on the books; they cease to be paid for those who are no longer on the books for any reason - including death.

In order to find out the current status of an individual, bona fide organisations such as the RSC can apply to the OPCS with details of date of death (if known). The OPCS through its National Health Service Central Register (NHSCR) can ascertain whether an individual

- is currently registered with a Family Practitioner Committee (FPC) (i.e. a patient on the books)

- is not currently registered with an FPC (e.g. has

moved out of the area and has not re-registered
with a fresh FPC)

- has left the country ('embarked')

- cannot be traced

- has died.

The OPCS can then furnish appropriate particulars, including,
where appropriate, a death certificate. This procedure is
time-consuming and if carried out repeatedly can become quite
expensive. In order to remove the necessity of repeated
requests for information about the current status of an individual
the NHSCR can place a 'flag' on his or her NHS records. When
there is a marked change in the status of the individual (e.g.
embarked or died) the interested organisation can be informed
accordingly. This is currently being put into operation for
members of the RSC.

At present, two retrospective mortality studies are being
conducted by the RSC with well-defined sets of individuals as
entry groups. One is in respect of 'the class of 1965' and
in the first instance traced male professional chemists with
an address for correspondence in England and Wales at 1st January
1965; it is now being expanded to include all professional
chemists in the UK and the Republic of Ireland at that date.
The other covers 'the class of 1980' and is tracing all
professional chemists with an address for correspondence in the
UK and the Republic of Ireland at 1st January 1980.

A striking case for such studies is made by reference to
two examples; one applies directly to the chemical industry
and the other to society as a whole.

As long ago as 1921, the chemical 2-naphthylamine was
highlighted as a suspected bladder carcinogen by the International
Labour Organisation; in 1954 in an exhaustive study of 455 cases
of bladder cancer in dyestuff workers in the British chemical
industry it was found that 311 were associated with exposure to

several compounds including 2-naphthylamine, for which compound
the risk of developing bladder cancer was 61 times that of the
general population with an average latency period of tumour
induction of 16 years [1];   in 1967 the presence, use, and
manufacture of 2-naphthylamine was banned in the UK. What
if the Institute of Chemistry of Great Britain and Ireland had,
in its earliest days, created a 'class of 1880' to track the
causes of mortality of its members? Perhaps many professional
chemists - including our own Eric Parker (formerly Registrar
and Secretary for Public Affairs of the RSC) - and associated
support staff would not have suffered a painful and degrading
death from bladder cancer, or from other occupationally
associated diseases.

The other example, which touches on society as a whole,
dates from four days in 1952. A dense fog covered the Greater
London area between 5th and 8th December in that year. A
strong impression was gained that the fog had produced severe
effects but it was only when mortality statistics in the form
of death certificates became available that the true extent of
the disaster emerged. In the week ending 13th December 1952,
the number of deaths from all causes occurring in Greater London
was 4703 compared with 1852 for the corresponding week in the
previous year.

Parliament passed the Clean Air Act in 1956 - a legislative
development which has done much to make 'pea soupers', 'smogs',
and 'London peculiars' anachronisms in the English language. In
fact the last London pea souper in easy recollection was in the
mid-1970s and it was claimed that this was due to involuntary
importation of products from the Ruhr.

Retrospective mortality studies on occupational or professional groups can provide pointers for more detailed
investigation. However, since they look back at dead people
and, to quote the 17th century proverb, 'dead men don't talk'
investigators are faced with many unanswered questions such as

- Did these people smoke a great deal?

- Did these people drink a lot?

-   Did these people work with particular chemicals?

-   Did these people have any family history of a given
    disease?

More detailed investigations can take the form of prospective
studies in general and prospective morbidity studies in particular.
The aim of prospective studies is, in most instances, to examine
the influence of a specific agent or factor on the incidence of
disease or death in a defined population. Prospective studies
are more demanding of time and resources than are retrospective
studies - but both call for an entry group e.g. 'the class of
a specified year'.

The entry group or 'sampling frame' can take several forms.
It can be composed of clerical workers in an organisation, every
tenth person on an electoral register, individuals on lists
maintained by Family Practitioner Committees, members of a trade
union, or professional people whose names are on a list maintained
by a co-ordinating body e.g. the General Medical Council or the
RSC.

A prospective study was mounted in 1951 among medical
practitioners in the UK and aimed at ascertaining any relationship
between their smoking habits and subsequent mortality. R. Doll
and A. B. Hill wrote to all practitioners on the medical register
in the UK and included a brief questionnaire for them to complete.
The fate of the group was then followed from that time. The
results derived in 1964 and reported on then[4] are shown in
the Figure. Some caution must be exercised in the interpretation
of these results: for example, some individuals may have reported
themselves as non-smokers having just given up smoking because of
the onset of disease; some may have presented an underestimate
or an overestimate of their habits.

A prospective study, well-known to most professional
chemists in the UK, is that which is currently being conducted
among some eighteen thousand members of the RSC who at 1st
January 1980 had an address for correspondence in the UK or in
the Republic of Ireland. The RSC Effects of Chemicals Assessment

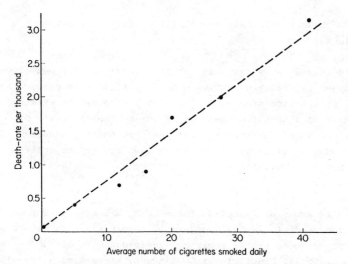

Death-rate from lung cancer, among men smoking different daily numbers of cigarettes at the start of a prospective study (men smoking pipes or cigars have been excluded).

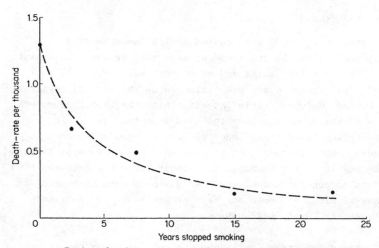

Death-rate from lung cancer, men continuing to smoke cigarettes and men who have given up smoking for different periods (men who had regularly smoked pipes or cigars as well as cigarettes have been excluded).

Figure From Doll and Hill [2]

Programme (RECAP) is a long-term investigation into the possible
health effects of chemicals on chemists. It operates by
tracking

- work characteristics (e.g.class of employment; type
of work; location of employment)

- social habits (e.g.smoking; alcohol, coffee, and
tea consumption)

- effects experienced (e.g. disturbance of the respiratory
system, dermatitis) which could be attributable to
chemicals

- use of pharmaceutical preparations (e.g. bronchospasm
agents, anti-allergy agents)

- history of ailments in the family

- history of work with chemicals.

It uses a detailed questionnaire completed at two-to-
three-yearly intervals by those professional chemists who are
regular participants in the investigation, the results of which
are computerised. As was mentioned earlier, at the same time
a 'flag' is being put on the NHS (not medical) records of the
whole of 'the class of 1980' so that the fate of participants
and non-participants alike can be followed. In addition the
'flag' makes it possible for the RSC to receive the death
certificates of those who ceased to be RECAP participants on re-
signing membership of the RSC.

In summary, at present the RSC is conducting two mortality
studies. One has some 17,000 people in the entry group (being
all professional chemists at 1st January 1965 with an address
for correspondence in the UK or the Republic of Ireland) and
is an expansion on the original group of some 15,000 male
professional chemists in England and Wales. The other has
some 24,000 people in the entry group (being all professional
chemists at 1st January 1980 with an address for correspondence
in the UK or the Republic of Ireland). The RSC is conducting

one morbidity study (the prospective RECAP investigation)
among some 18,000 professional chemists who, on an entirely
voluntary and altruistic basis, have given extensive details
of their work pattern and social habits. Can these mortality
and morbidity studies highlight anything? One mortality
study is reasonably self-contained, since the 15,000 males
were traced over a 15 year period from 1965 to 1979. It has
revealed that, in round figures, over the 15 years some 2,000
have died, some 1,000 have left the UK, and some 12,000
(including myself) have come through, more or less unscathed.
One person (0.06%) of the group) has not yet been traced.
Since I am told that 100% successes are viewed with suspicion
by epidemiologists we have at this 99.94% level called off our
sleuths, who to reduce the number of no traces to this modest
level have used not only the OPCS but also electoral lists,
telephone books, and advice from personnel departments

      Some analyses have been carried out on a basis of
Proportional Mortality Ratios (PMRs) and of Standardised
Mortality Ratios (SMRs). Preliminary results indicate that,
although chemists seem to be a relatively healthy lot, there
are pockets of excess deaths and some pointers for more
detailed inspection in the RECAP investigation.

      RECAP itself is a relatively youthful specimen. Its
participants are telling us that they are modest in their
overall smoking habits and are becoming even more modest.
Their alcohol consumption seems also to be at a modest level.
All in all they ought to be good insurance risks .

      But remembering the quote attributed to John Maynard
Keynes 'in the long run we are all dead' - and regrettable
as it may seem - we have to await a build-up of death
statistics from the 18,000 in the class of 1980 to be assured
of the total contribution from this particular prospective
study. Pointers from the 1965 mortality study lead us to
believe that by 1985 we shall have around 750 or more and by
1990 in excess of 1,500 deaths to contribute to the investigation.

      Morbidity and mortality studies have been carried out on

chemists in other parts of the world and include, in the
USA a mortality study on those who had been in membership
of the American Chemical Society [3] and a morbidity and
mortality study on those employed by E. I. du Pont de Nemours
and Co. Inc.; [4] in Sweden a morbidity and mortality study on
graduates of the Royal Institute of Technology, Stockholm.[5]
To the best of our knowledge the RSC studies are the most
comprehensive of those that have been, or are being, carried
out.

So that we may concentrate on the symposium itself may
I remind you of the title of this session: 'Health and
Safety in the Chemical Laboratory - at what cost?' The cost
of running one of our studies is around £30,000 and on top
of this has to be added an equal sum for the staff costs.
Put another way if the RSC were paying directly for a study
it would be at the rate of £2 per professional member for
each study. However, since these studies are seen by the
Commission of the European Communities (CEC) to have relevance
to the European Community as a whole, the CEC is providing
support funding amounting to half the cost of a study. But
there are other kinds of cost. The OPCS in its publication
'Mortality statistics 1980' [6] estimates that for people aged
between 15 and 64, 1.9 million years of working life are lost
as a result of early death. There will be a cost in terms
of loss of output (especially noticeable in highly trained
individuals such as professional chemists). There will also
be a loss to the family in terms of loss of earnings and loss
of companionship.

We know that a proportion of these early deaths are
occupationally based, and almost certainly individual
participants at this symposium will know of somebody who
has died in circumstances where there is a suspicion that
occupational exposure to certain chemicals was a contributing
factor. Philip Morgan and Neil Davies [7] have looked at costs
of occupational accidents and diseases from an objective
standpoint (resource costs: including costs of lost output,
damage to plant and machinery, medical treatment, administration) and from a subjective standpoint (pain, grief,

and suffering). They have arrived at a total cost for
1978-1979 of £90 million.

If we accept that there is a case for the conduct of
morbidity and mortality studies - and we have the example
of 2-naphthylamine and bladder cancer to make the case -
then the cost is relatively small.

The principal object of the RSC as set out in the Royal
Charter is 'the general advancement of chemical science'
and for that purpose it is called on 'to serve the public
interest by acting in an advisory, consultative, or
representative capacity in matters relating to the science
and practice of chemistry'. Thus the RSC, based in and on
the United Kingdom, is uniquely equipped to provide a service
to society as a whole in this realm. Information from the
studies, as it becomes available over time, will be of value
not only to professional chemists but also to their support
staff. Earlier in this paper mention was made that one
'sample frame' was members of a trade union. It would surely
be a positive contribution by the trade union movement in
this country to its members if it were to initiate a programme
of studies similar to those described in this paper. I leave
you with that thought.

Since this is an international symposium may I close with
a quote from Nikita Khrushchev:

'Life is short, live it up'.

References

1.   R. A M. Case, M . E. Hosker, D. B. McDonald and
     J. T. Pearson, (1954). 'Tumours of the urinary
     bladder in workmen engaged in the manufacture and
     use of certain dyestuff intermediates in the British
     Industry; 'The role of aniline, benzidine, alpha-
     naphthylamine and beta-naphthylamine', Br. J. Ind.
     Med. 1954, 11, 75.

2.   R. Doll, and A. B. Hill, 'Mortality in relation to
     smoking; ten year's observation of British Doctors'
     Br. Med. J. 1964, 1, 1399.

3.   F. P. Li, J. F. Fraumeni, Jr., N. Mantel and R. W.
     Miller, 'Cancer Mortality among Chemists', J. Natl. Cancer
     Inst., 1969, 43, 1159.

4.   S. K. Hoar and S. Pell, 'A Retrospective Cohort, Study
     of Mortality and Cancer Incidence among Chemists',
     J. Occup. Med. 1981, 23, 485.

5.   G. R. Olin and A. Ahlbom, 'The Cancer mortality among
     Swedish chemists graduated during three decades',
     Environ. Res. 1980, 22, 154.

6.   Office of Population Censuses and Surveys, Mortality
     Statistics, Series DH1, No. 9, 1980, p. 17, Table 24,
     Her Majesty's Stationery Office, London.

7.   P. Morgan and N. Davies, 'Costs of occupational accidents
     and diseases in Great Britain', Employment Gazette,
     November 1981, 477.

Discussion

Q1.   I would like to congratulate Mr Henman on his
        lucid account of the RSC morbidity and mortality
        studies – a pleasant contrast to many papers on
        epidemiology that I have listened to! I would
        like to make one observation and raise one
        question:

            Studies like those described are expensive
            but in my opinion the costs are trivial
            compared to the suffering that may occur
            and to the litigation costs which may be
            incurred.

When the results of these studies are
published it will not be sufficient merely
to say that a participant is embarked. It
will be necessary to say whether he is dead
or alive. I imagine that when such data are
deducted the studies will be down to a more
conventional 97% or so trace. Is this so?

Henman - Yes, that is true.

Q2.   The RSC work is seen by other European Chemical
           Societies as a good example of how to perform such
           epidemiological studies. I remember that at a
           meeting of the Working Party on Professional Affairs
           of the Federation of European Chemical Societies,
           Mr Henman said that the average death rate among
           chemists is not higher than that for the rest of
           the population - although the causes of death may
           differ from the 'average'. If this is correct I
           think it is important for the image of chemists,
           chemistry and the chemical industry that it should
           be widely appreciated.

Henman - That is correct. There are some considerable
           underscorings from certain causes of death but
           there are some overscorings as well. Clearly
           there are occupationally related diseases but if
           we can focus on these then chemists will be an
           even healthier lot! Chemists are also pretty
           healthy when compared with other professional
           groups. It may be that we do not insult ourselves
           with over indulgences quite as much as others do?

Q3.   Mr Henman mentioned the case of 2-naphthylamine
           and the long period between recognition of a
           problem and the passing of legislation. In the
           UK we still have only the Carcinogenic Substances
           Regulations but in recent years several substances

not covered by these regulations have been
recognised as human carcinogens. When can we
expect new legislation to deal with such materials?

Henman -   I am perhaps not the best person of those present
to answer that question.

There is a directive in the European 'pipeline'
which may take some 3 years to come through and
which is relevant to this question. It will
mirror our own Carcinogenic Substances Regulations
but will make some provision for 'layering-on'
extra chemicals.

I am afraid that I don't know whether the UK is
planning to supplement the 1967 regulations. I
would defer to Mr Luxon on that matter.

Luxon -   This is a difficult problem. The question really
is 'what is a carcinogenic substance?!' Until
that is sorted out it is going to be very difficult
to proceed.

The 1967 regulations were directed at a specific
industry - the dyestuffs industry. It is very
attractive to extend the regulations but you
are immediately beset with the problems of
defining what is a carcinogen. This is difficult
to decide both nationally and internationally.

The matter is under consideration in the HSE
but what we really need is an internationally
accepted definition of a carcinogen.

Henman -   Perhaps that could be one recommendation to
come out of this symposium?

Economics of Health and Safety Measures

By Dr D. van der Steen

HEAD OF SAFETY DEPARTMENT, UNILEVER RESEARCH LABORATORY, VLAARDINGEN, THE NETHERLANDS

Introduction

With regard to health and safety measures it is, in my view, very difficult indeed to do a cost-benefit analysis. The literature gives few references and those that are of any value are sometimes so complicated that I wonder whether it is justified to do a cost-benefit analysis after all. The complex of problems associated with such an analysis has much to do with that of a risk analysis. Such an analysis relates, in many cases, to a certain event which may affect the environment. Here, the implementation of certain measures is taken into account, which may decrease a certain risk.

In risk analysis, the consequences of a certain event are estimated. To determine a certain risk, the chance of such a risk happening is also taken into account. This implies that an event, which is not very likely to occur but of which the consequences may be far-reaching, will result in a similar risk as an event which is very likely to take place but of which the consequences will be small. In risk analysis, a limit is often put on the accidents to be considered. This limit is frequently the number of people one may expect to be killed. In some analyses certain recommendations are made, so that it is suggested that the activity concerned is safe enough. Numerous examples show that there are great objections to such a starting point - both from experts and from certain population groups.

Against this background, one may ask whether the population (or parts of the population, who are really the core of the problem) is only interested in fatal accidents or whether it is interested in other matters as well. Among such matters may be:

- Long-term effects;

- The possibility that a certain region will be impossible to live in for a short or longer period of time;

- Sudden threat to human health.

These are examples of questions that are vital to many people but about which few details are known.

Although risk analysis may be very important, especially in cases where risks are compared, I must, I think, put some question marks against analyses with a large number of built-in uncertainties. The number of such question marks becomes even larger if the cost factors are taken into account. In such analyses, when expressing the probability of fatal accidents occurring, an economic value is put on an individual - but many are of the opinion that man is more than an economic asset. In this light, it is logical that the general population should become more and more interested and involved in estimated risks. In my view, we must therefore consider the following aspects in regard to such an analysis:

uncertainties should be removed as far as possible;

the general public may interpret the analysis in its own way;

an answer should be given to the question of the way the damage done (which cannot be expressed in terms of money) has been designated.

This complex of problems has also played a role in the considerations regarding the cost-benefit analysis of health and safety measures.

Approach of a Cost-Benefit Analysis

It is well known that a risk diminishes with increasing prevention costs. In his publication[1] Craig Sinclair gives a clear picture of this. A graph (Figure 1) may be an even better tool to

illustrate it; here the cost of accidents is plotted against
the prevention costs. This shows that the accident costs
decrease (the risk is diminished at the same time) whereas the

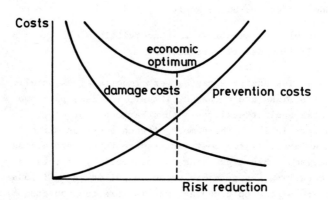

Figure 1   Economic optimum between damage and prevention costs

prevention costs increase. To reduce the cost of accidents as
much as possible, the prevention costs will, of necessity, be
high. With the help of an economic optimum (given by the author),
it is possible to show that the costs connected with accidents
will increase as the prevention efforts are increased. In fact,
this is not very revealing as long as there is no clear indication
what the reduction of the risk means and in what way it can be
measured.

In many cases the risk costs are clearly defined. We know:

the cost connected with medical treatment;

the cost caused by damage to property;

the administrative costs, which are the costs connected
with the investigation into the accident;

the cost connected with production losses. It should be

noted here that in the case of a non-fatal accident,
there is no doubt a visible damage for the employer.
However, in the case of a fatal accident, (particularly
in cases where employers are not to blame for the
accident), the State is in fact to blame and not the
employer, for in such cases the State becomes responsible
for the relatives of the victim.

These are measureable costs but there are, of course, also costs
which cannot be measured - for example the life of an individual.
Closely connected with this question is the fact that we may
wonder whether we have a right to make a pronouncement here.

A recent investigation by two Dutch economists throws some
more light on this question. In 1982 Van den Bosch and
Petersen[2] investigated the extent to which there is a relationship
between regulations regarding health-care and the health-aspect
correlated loss of income regulations. They concluded that
health-care in the Netherlands falls into two categories:

curative health care, which is that part of the
health-care directed towards the cure of certain
illnesses;

preventive health-care, which is primarily directed
towards the prevention of a worsening of a state of
health.

According to the authors, their findings showed that as health-
care is expanded the positive influence of additional health-care
facilities on the state of health decreases. They even postulate
that above a certain limit too great an intensification of
health-care may, on balance, have a negative effect on the state
of health. Again, according to these authors, above a certain
degree of health-care, there is the chance that the more
intensive the health-care becomes the less healthy one feels.

Figure 2 shows the relationship between the costs connected
with an appeal to the Industrial Disability Law and the costs
connected with health-care. Extension of the health-care

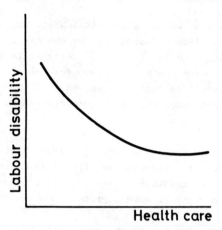

Figure 2   <u>Health care and labour disability</u>

facilities has, as a rule, a stabilising influence on becoming
unemployable. This implies that, with increasing health-care
facilities, appeals to the Industrial Disability Law payments
decrease. At a certain extent of health-care facilities the
reverse is observed.

    The authors do not even exclude the possibility that the
minimum has been passed and that a further extension of
health-care facilities might lead to more appeals to the
Industrial Disability Regulations. Also in this case we might
speak of an economic optimum - as Craig Sinclair[1] does. This
author compared the accident costs with the prevention costs
but Van den Bosch and Petersen compared the payments made by
virtue of the Industrial Disability Law with the costs of all
the health-care facilities.

    Although there are certain agreements, there are also
differences between these authors. The most striking difference
is that in Sinclair's graph, the prevention costs are plotted
whereas in the other the total health-care costs are shown.

    Despite the fact that it is not always easy to draw a line
between the two areas, Van den Bosch and Petersen postulate
that the costs of preventive health-care over the period from

1968 to 1981 were <u>ca</u>. 3% of the total health-care costs.
Within the framework of preventive health-care, about one-third
of the total means went on preventive industrial health-care.
The costs of industrial health-care are, therefore, 1% of the
total health-care costs. In addition to the fact that this
percentage is low there is another point of interest to note.
If we look again at the graph constructed by Van den Bosch and
Petersen, and plot on the <u>x</u>-axis <u>not</u> the total costs of
health-care but those of preventive care, the course of the
curve may become quite different; the values on the <u>x</u>-axis
will then be decreased by one-hundredth. There is a chance that
the position of the economic optimum (if there is one) becomes
a debatable point and that an increase in prevention costs will
cause a (drastic) decrease in the number of appeals for benefit
payments. This assumption is in agreement with the findings of
Van den Bosch and Petersen. They showed that an occupational
health centre tends to decrease costs. On the basis of the
outcome of careful calculations, they arrived at the conclusion
that if dfl 600 million were to be spent on (preventive)
occupational health centres, this would save dfl 1400 million
on benefit payments, giving a nett benefit of dfl 800 million.

It should be noted, however, that the work I have just
mentioned covers the field of health-care and that many
preventive measures taken in industry in relation to health and
safety are complementary to it. Against this background
therefore it is, in my view, not incorrect to assume that the
complex of problems here is of the same order.

According to various authors, the cost of preventive
measures is related to an optimum, although one might be tempted
to put some question marks here.

Accidents in Different Branches of Industry
As you will have noted, I have put quite a number of question
marks against the available cost-benefit analyses. Despite
this I will try to answer the question of 'how to proceed' and
attention will be given mainly to industrial enterprises and
laboratories. I will present a different approach which I
think may be useful.

In the first place, I tried to establish whether there
are differences between accidents - fatal or non-fatal - in
different branches of industry in different countries. With
regard to the types of work, I confined myself to the work
done in agriculture, in construction and in other industries.
I did so because it is in these fields that most accidents
occur. With regard to different countries I tried in the
first place to compare the data for some EEC countries with
those of countries outside the community - including some
developing countries. I must say in fairness that my choice
was greatly determined by data contained in the Year Books of
Labour Statistics of the International Labour Office[3];   I would
have liked to include the UK as well, but have not succeeded in
finding any relevant data. Figure 3 indicates the number of
accidents per 1000 staff during the period 1975 to 1980 for
agriculture, construction and other industries. You will see

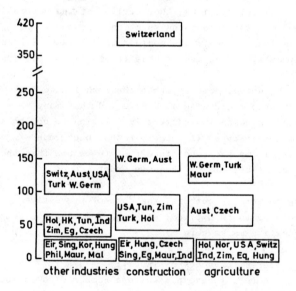

Figure 3   <u>Number of accidents per 1000 employees 1975-1980</u>

that the number of accidents for the different branches of
industry is quite different and is highest in construction and

that industry as a whole shows a fairly good picture also when
compared with agriculture. The graph is an overall picture
of the situation but the figures for Germany, Austria,
Switzerland, and Turkey are rather high.

If we look at the number of fatal accidents per 1000
employees, we see roughly the same picture (Figure 4). Also in
this case, the number is highest in construction and lowest in
other industries as a whole. Germany, Austria, and Turkey are
again rather high.

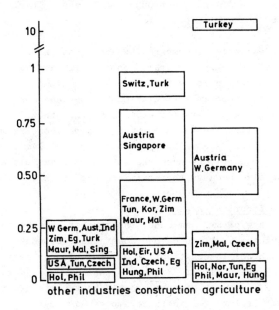

Figure 4   Number of fatal accidents per 1000 employees 1975-1980

What makes industry in general a 'safe workplace'? I would
like to mention a few points, which do not yet often apply to
construction and agriculture:

in general, Government has geared legislation more to
industry than to construction or agriculture;

in contrast to activities in construction and agriculture,
the work in industry is mostly done at one location so

that it is easier for a factory management or Government
to exercise control;

the influence of safety committees and other bodies
having regulatory power is highest in industry; as a
result, this influence certainly has positive
consequences;

the cost of accidents in industry is higher than in
other branches of industry because such accidents often
mean damage to equipment and production losses. This
aspect may be the most important of all.

It would be desirable to compare the accidents in the
various branches of industry (general, construction and agriculture)
with the costs (production losses and damage to equipment) and
to establish whether there is any relationship between costs
on the one hand and the number of accidents on the other.
Although I have not found any data on such a comparison, we may
assume that laboratories - as a part of industry - will give an
even better picture, partly because the activities take place on
a small scale but also because the staff are highly skilled.

Are Accidents the Important Aspect?
Although the data are clear I am under the impression that this
is not the true picture. It is beyond doubt that in industry
generally, although not in construction or to a lesser extent
in agriculture, we are confronted with professional diseases.
According to Dutch Law, such diseases should be reported to the
bodies concerned. Ireland for instance imposes a similar
obligation but an Irish report[4] says that such diseases are
seldom reported - although doctors are obliged to do so. In
many cases this is not a question of unwillingness but rather
a consequence of the fact that it is often extremely difficult
to relate the disease to the profession. For this reason the
cost connected with professional diseases will show up
insufficiently in a cost-benefit analysis. However, it is not
only the professional diseases in the cost-benefit analysis which
distort the picture. There are many other factors that affect
such an analysis. I would like to mention just one.

At the moment factories are shut down or threatened with closure as a result of the economic situation. This threat and the feeling of being sidetracked no doubt affect the state of health of employees. Apparently, this has nothing to do with the current measures taken in the field of health and safety. However, if such closures do influence the state of health of employees the cost-benefit analysis becomes even more complicated. In a recently published Dutch report[5] some reasons are given for the decreasing sickness absence in areas with a high unemployment figure. One often hears that an explanation for this might be the dismissal of less healthy staff and the threat of losing one's job in the case of a high rate of sickness absence. But the threat of losing one's job may also create a stress situation and thus increase the sickness absence figures.

Recently the Dutch Government ordered an investigation into the possibility of measuring the prevention of stress. The reason behind this is that branches of industry are increasingly confronted by health complaints such as head-aches, stress, and depression, which are possibly caused by the organisation of the work, the work-load, etc.

At the beginning of my talk I said that it is a difficult job to carry out a cost-benefit analysis. I also tried to indicate that the cost relating to accidents is slight for laboratories. However, I have also tried to indicate that safety in industry and in laboratories is frequently interpreted on the basis of accident statistics and that statistical data on professional illnesses are not adequate.

What else can we do in regard to health and safety measures? It may be useful to point to some recent developments in different countries.

Developments on a National Level

Safety and health are increasingly integrated in the policy of a company and thus constitute part of the job. This means that, from the beginning, safety and health aspects should be taken into account in the production and handling of dangerous substances and machines.

Employers and employees are to an increasing degree
forced to work together in matters of safety and health, for the
major responsibility to decrease the number of accidents and
professional illnesses rests with those who create certain risks
and with those who have to accept these risks.

Developments on an International Level

As already happens in some countries, the social partners
should become involved in the formulation of norms and
standards. There is a tendency for the field of health and
safety to be broadened - for instance the humanisation of work.
The rapid developments in technology and science have made some
countries decide to restrict the formulation of specific
standards as much as possible and to oblige employers to create
a safe workplace. Such developments are already in progress in
Germany, the Netherlands, Denmark, the UK and Italy.

Thus it is up to the employers to decide on the means of
realising the objectives within the framework prescribed by Law.
Different countries have different laws but there are also
aspects that are in agreement. Some trends are clear:

the co-operation between employers and employees.
If co-operation in the field of safety and health
between these partners is to be optimal both partners
should have the same information at their disposal - one
possibility is for experts (the medical and safety
officers) to be at the disposal of both;

the abandonment of differing standards; at the same time,
employers should be charged to look after the safety and
health of employees.

The consequences of such regulation will, in my opinion, be
far-reaching. As you know, the Government cannot always react
quickly to situations. Now the employer is in a position to
react to such a situation without being hindered by 'slow'
decisions. This has certain advantages for the employer, but
also consequences. Allow me to give you some examples.

Shift from Stringent to Mild in Regulations.
When DNA recombinant work was started it was researchers in
the United States in particular who pointed to certain possible
risks connected with this type of work. The consequence of
this was that relevant regulations were published in the USA -
and also in the UK. While in Europe, and in particular in the
Netherlands, discussions about the possible dangers of such
work were still going on the USA was formulating much milder
regulations in regard to DNA recombinant work because, it was
said, the dangers had been exaggerated. A short time afterwards,
the regulations also became much less stringent in the
Netherlands.

Such a rapid adaptation, which offers many advantages, is
only possible within the framework of existing legislation,
and the bodies involved have considerable responsibility here.
It is also clear that in many areas a rapid adaptation to the
regulations in force can be realised. In this case, the
adaptation to less rigid regulations went smoothly and without
giving rise to any difficulties.

'Possible' Shift from Mild to Stringent in Regulations.
On the other hand, employers may be obliged to conform to more
stringent rules. In this connection, I would like to mention
the work of Sëppalainen,[6] in which she says that workers working
with organic solvents would show an abnormal electroencephalographic
pattern.

Lindström says in his publication[7] that the most severe
decreases in behavioural functions were found among workers with
occupational disease caused by aromatic and halogenated
hydrocarbons and their mixtures. He also pointed to a possible
synergistic effect of various solvents.

A publication by Mikkelsen is also worth noting,[8] in which it
is said that there is an increased risk of presenile dementia
among painters.

In many countries, threshold values have been fixed for
many substances including solvents. Studies of the synergistic
effects of certain solvents have been performed and I wonder

whether it is not one of the first tasks of employers to
investigate this problem further and adapt existing regulations
to more rigid ones, where necessary.

Quite a different problem, which has become of interest
of late, is that connected with visual display units. The
International Labour Organisation (ILO) has said that many
people complain about these display units giving them head-ache,
back-ache, and strained eyes. According to the ILO these
complaints are caused by the high frequency of looking up and
down from the VDU to the keyboard. The organisation recommends
not working with VDUs for longer than 4 hours.

Digernes and Astrup[9] looked at quite a different aspect
of these display units. They investigated whether skin rashes
and eczema would be found in people working with these units
as a result of contact with PCBs from leaking units. In rooms
housing VDU equipment the authors found a clear increase in
the PCB concentration although these concentrations remained
below the safety limit recommended by NIOSH in 1977.

However, in this case the possible risk from a combination
of substances, the concentration of each of which is below
the danger limit, should be investigated further.

Finally I would like to ask some questions and draw some
conclusions:

-   Are we really heading in the right direction if
    we weigh costs against benefits? Should we not
    regard the taking of measures as an investment
    rather than something that only increases costs?;

-   In the future a sound cost-benefit analysis will
    no doubt be carried out with respect to health
    and safety measures. In this case I consider it
    recommendable that such an analysis be made by a
    team including economists and safety advisers;

-   It is well-known that accident statistics for
    laboratories do not tell a lot but that statistics

for occupational illnesses tell a lot more.
Because the relationship between job and
occupational illness is not always easy to establish
- if done at all - other routes would have to be
explored to allow the establishment of such a
relation. A simple procedure might be the inclusion
of the health consequences of the job in the
statistics of laboratories;

- Although much research work is already done of the
dangers of toxic compounds and efforts are being
made to extrapolate the results of experiments to
man, I am nevertheless under the impression that
the investigations might also be directed to solvents
and to the combinations of materials used in chemical
laboratories;

- The measures to be applied in the field of safety
and health should be taken on a sound basis using the
latest scientific findings.

References

1. T. Craig Sinclair, 'A Cost-Effectiveness Approach to
Industrial Safety', Her Majesty's Stationery Office,
London, 1972.

2. F.A.J. Van den Bosch and C. Petersen, 'Gezondheidszorg en
arbeidsongeschiktheid', Openbare Uitgaven 1982/5a en blz. 37.

3. 'Year book of Labour Statistics 1981', International Labour
Office, Geneva.

4. Veiligheid en Gezondheid op de werkplek in de Europese
Gemeenschappen, 1980.

5. Nederlands Instituut voor Preventieve Gezondheidszorg TNO.

6. A.M. Sëppalainen, <u>Scand. J. Work Environ. Health</u>, 1981, 7, 29.

7. K. Lindström, <u>Scand. J. Work Environ. Health</u>, 1981, 7, 48.

8. S. Mikkelsen, Scand. J. Soc. Med., 1980, 16 (Supplement), 34.

9. V. Digernes and E.G. Astrup, Int. Arch. Occup. Environ.
    Health, 1982, 49, 193.

Discussion

Q1.   Do you feel that cost/benefit analyses for
               health and safety measures should be carried
               out jointly by unions and management?

Van der Steen - Yes, I do feel that they should.

Procedures and Statistics in France

By Dr C. Mordini
RESEARCH DIRECTORATE, RHÔNE-POULENC RECHERCHES, COURBEVOIE, FRANCE

At the present time France has a working population of
21 million people. Of these 14 million work in all kinds
of industry. The remaining 7 million belong to the army,
administration, agriculture, etc.

All industrial activities are officially distributed
in fifteen groups. The chemical industry with 350,000
people is one of the fifteen.

The groups are supervised by four ministries:
Research and Industry, Labour, Health, and Environment.

Industrial activities are subject to four main groups
of regulations concerning:

- Labour (laws of 12/1976, 8/1982, and 12/1982);

- Social insurances (cost of accidents, allowances
  and assets, controlling exposure);

- Industry (listed workshops, protection of environment,
  transport of products, and labelling);

- EEC directives (sometimes difficult to interpret and
  to apply).

The government controls the application of regulations
by civil servants:

- Labour Inspector, who depends on the Ministry of
  Labour (with considerable authority);

- Social Insurance Inspector (dependent on the
  National Social Insurance Administration;
  also an adviser for manufacturers);

- Listed Workshop Inspector.

In each factory, a commission called 'Health, Safety and
Environmental Work Conditions' must exist. This is composed
of the Manager, President, representatives of the workers and
experts, such as the doctor.

Safety Policy
The improvement of safety results cannot be achieved without
some basic conditions:

- Commitment of the general management and of the
  Chairman himself, and other levels of responsibility;

- Taking into account safety in the activity, concerning
  people (knowledge and training) and structures (safety
  audit and investment in process equipment and products);

- Identification and analysis of hazards (faultry tree
  method);

- Having safety officers of high potential level able to
  have personal authority.

These goals will not be reached by technical means alone,
(important as these are): Communication is much more important -

safety = communication

Today the Safety Officer has to be more a man of
communication than a technical scientist.

Accidents Specific to the Chemical Industry
The most frequently observed concern:-

- Internal cleaning of equipment (tank, reactor);

- Asphyxia with nitrogen;

- Toxic gas;

- Degassing purges, pipes;

- Loading and unloading stations;

- Whirling machinery.

The chemists working in laboratories belong to:

- University and national scientific organisations;

- Industrial research centres;

- Laboratories controlling properties of chemicals in factories.

It is very difficult to know the specific statistics for such chemists because:

- Universities do not publish their results. Only since 1982 have they been obliged to appoint a Safety Officer and Safety Commission;

- Laboratories of chemical control are totally included in factories;

- Only certain research centres publish their statistics (Rhône-Poulenc in France for example).

In résumé, we can say that:

- About 80% of accidents in laboratories are not specific;

- Chemists are less sensitive to safety and feel that hazards are less dangerous;

- Chemists feel that they are more able to prevent accidents.

How to Present the Safety Results
Several methods are used:

- Number of accidents which stopped work;

- Number of lost days;

- Number of hours worked;

- Permanent disability rate;

- Temporary disability allowances;

- Permanent disability allowances (really assets).

The following official rates are calculated:

- Frequency rate (number of accidents with stopped
                 work per million hours worked);

- Severity rate (number of lost days per thousand hours
                 worked);

- Severity index (amount of permanent disability rate
                 per million hours worked);

- Hazard index   (ratio of all allowances and assets to
                 the total payroll).

Results

Figure 1 presents a comparison between the chemical industry and the whole of industry over 25 years. The frequency rate decreased drastically from 1955 to 1973; since 1973 the decrease has been much less and presents a plateau. The number of lost days in 1980 for the chemical industry is equivalent to a factory of 2,500 people closed for a year.

In 1981 there was a reduction of 6% with 510,000 lost days.

Figure 2 gives some further details for 1980. The average cost of an accident in the chemical industry is the most important after that of one in the construction industry. The total cost for 1980 was F 243 million, which represents 1.3% of the total payroll.

Figure 3* shows the variations of the frequency rate for the fifteen groups of French Industry over 25 years from 1955. We observe a clear break in 1973.

Figure 4* reveals a continuous increase in the average duration of stopped work. This reached 28 days in 1980. Several causes may explain this situation and these are now studied.

Figure 5* shows the analysis of the average time of stopped work according to age, and Figure 6* shows the results of the main types of accidents for the whole set of activities: circulation, (walking, going up and down stairs), fall, handling, transport of things by hand, accidentally moving objects, raising work, vehicles, machines, tools. We observe that circulation and handling are responsible for 50% of the frequency rate.

In Figure 7* the same results for the chemical industry are presented. The three first groups represent 60% of accidents and for laboratories this percentage reaches about 75%.

Statistical analysis of the areas of the body injured shows the following results for the chemical industry: head 5%, hand 26% and trunk 18% of total accidents. For the whole industry accidents to the hands represent 32% of the total.

Footnote: * From publications of the 'Caisse Nationale de l'Assurance Maladie des travailleurs salariés'.

		1955	1965	1973	1975	1977	1979	1980
T. F.	WHOLE	53	47	40	40	37,5	35,7	34,9
	CHEMISTRY	51	38	32	32	31	30	29
T. G.	WHOLE	1,09	1,09	1,03	1,11	1,04	1,00	0,98
	CHEMISTRY	1,08	0,94	0,85	0,89	0,85	0,83	0,81
I. G.	WHOLE	59,7	56,4	46,7	46,7	42,9	38,2	35,8
	CHEMISTRY	53,3	54,0	40,2	38,1	38,1	32,4	30,6
J. P.	WHOLE							27270000
	CHEMISTRY							545530 *

* equivalent to a plant of 2.500 peoples closed for a year.

T F: frequency rate.

T G: severity rate.

J P: lost day

Figure 1

1978 - 1979 - 1980	W H O L E	CHEMISTRY (1980)
WORKING PEOPLE	14 000 000	350 000
TOTAL ACCIDENTS	970 000	19 370
ACCIDENTS WITH PERMANENT INCAPACITY (I.P.)	102 000	2 300 (12%)
D E A T H	1 465	39
RISK INDEX	1,65%	1,32%
AVERAGE COST OF ACCIDENT	FF. 3 650	FF. 4 200
AVERAGE COST OF ACCIDENT WITH I.P.	FF. 50 300	FF. 57 000

Figure 2

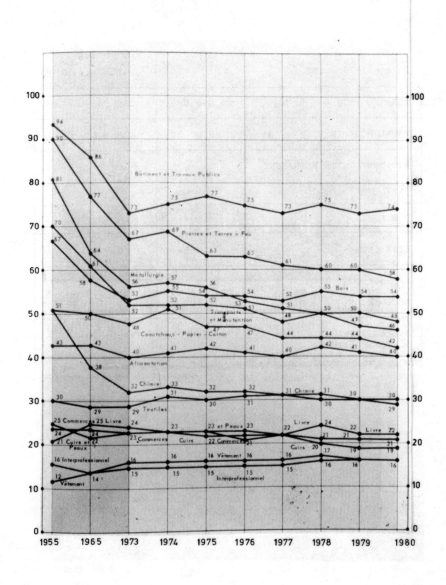

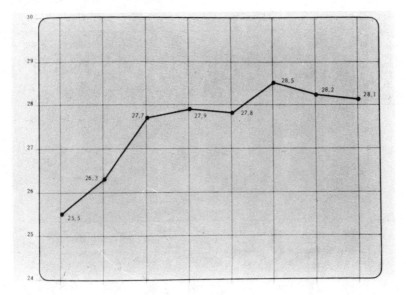

Figure 4

Tranches d'âge	Durée moyenne de l'incapacité temporaire (en jours)	Taux moyen de l'incapacité permanente (en %)
— de 16 ans	19	10
16 et 17 ans	17	9
18 et 19 ans	17	8
20 à 29 ans	21	9
30 à 39 ans	30	9
40 à 49 ans	35	10
50 à 59 ans	38	11
60 à 64 ans	41	11
65 ans et plus	45	14
Moyenne	**28**	**10**

Figure 5

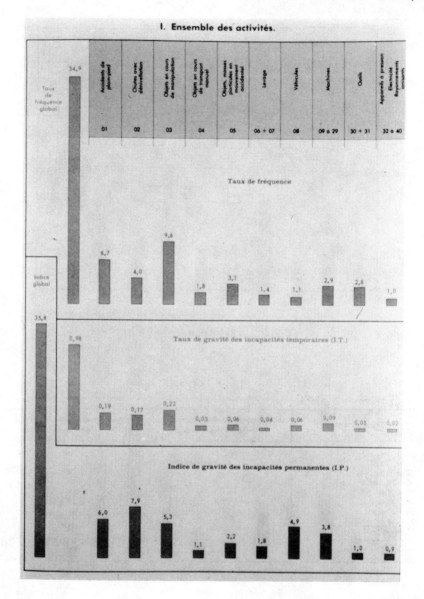

Figure 6

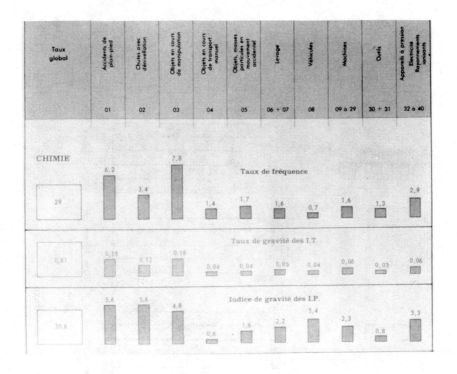

Figure 7

Figure 8* shows the breakdown of injured people according to their age. We observe that the percentage of accidents is more important among workers who are less than 30 years old; however, the severity increases if the workers are older.

Finally Figure 9 gives a global analysis of accidents according to six criteria, both for the whole industry and for the chemical industry, which behaves differently for the three first criteria.

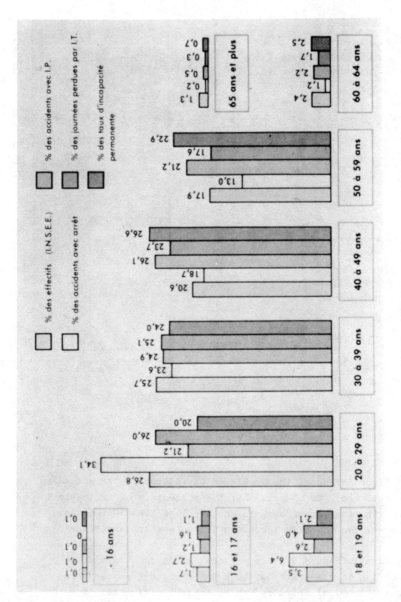

Figure 8

DATA ANALYSIS

Accidents with stop.		Accidents with permanent incapacity		Working lost days	
Whole %	Chemistry %	Whole %	Chemistry %	Whole %	Chemistry %
PEOPLE AGE					
<20 9	3	4	1,5	6	2
20–40 57,5	53	46	38,5	51	46
>40 32,5	44	50	60	43	52
NATIONALITY					
French 81	88	78	86	76	86
Foreigner 19	12	22	14	24	14
SEX					
Men 87	85	87	85	72	86
Women 13	15	13	15	28	14
NATURE OF INJURIES					
Wound + Bruise 62	59	49	50	54	53
Other 38	41	51	50	46	47
PLACES OF INJURIES					
Hands + Breast + Low members 63	61	59	58	62	61
Eyes 5,5	5,5	2	2,5	2	2
Other 31,5	33,5	39	39,5	36	37
PHYSICAL EVENTS					
Circulation + handling 57	60	57	59	60	60
Other 43	40	43	41	40	40

Figure 9

Conclusion

The results stated show that important progress has been achieved but much should still be done and we cannot be satisfied.

Accident is primarily a misfortune for the person involved, but it also costs society more and more money. (One can estimate that the indirect cost of an accident is 2.5 to 4 times the direct cost (taxes paid by the company to the social insurances). Today the probability of an accident occurring depends more than before on the psychological environment in the company and in the workplace. Unemployment, uncertainty about the future and lack of communication are new factors which should be taken into account. Therefore we cannot expect improvement in health and safety from techniques only but more from the ability of managers and other people with responsibilities, to communicate, to understand, and to include safety in their technical goals.

Discussion

Q1.   You mentioned 'safety audits'. Do you apply these in industry and universities? How often do they take place? Are they pre-arranged or unannounced?

Mordini - We have two types of safety audit -

          (i) Safety audit for investment. Any proposals for technical investment must be seen by the people with responsibility for safety matters. This is to ensure that safety considerations are not forgotten when making technical investments.

(ii) The second type is used mainly in research establishments. These tend to be more clearly defined entities with well trained staff who are able to assist with the audit. A safety officer from one plant performs an audit on another plant. This occurs once a year.

Q2.   Do such safety audits apply in university laboratories? If not do you think it would be a good idea if they did?

Mordini - No, they do not apply in university laboratories. I think it is too early for that.

Universities still have to absorb new regulations. They need to improve generally because a safety audit requires that there is already a good level of safety awareness and technique.

Q3.   It seems to me that chemists working in laboratories often have the naïve attitude that they are safer than other people. Do you have any ideas on how to change this attitude?

Mordini - Perhaps this is the main problem that we face - changing people's attitudes. The main problems that face us are not technical but are concerned with communication. It is essential that top management (the managing director in industry or the Chancellor in a university) takes safety matters fully into account. Without this we will not get the results we require. We also have to make people more sensitive to the financial costs of accidents - not only to the company concerned but also to the rest of society.

Q4.   One of the factors that has made managers very
             sensitive to safety matters in the USA is the
             threat of litigation for negligence. This is
             especially so in the case of long-term effects
             of substances where several cases are pending.
             Is there anything comparable in France, where
             the worker sues the company directly for damages?

Mordini - We have regulations that make the management
          responsible for accidents.

Q5.   How do you define a 'factory' for the purposes
             of French Safety Law?

Mordini - 'Factory' is the wrong word. The law applies
          to any organisation with a minimum of ten
          workers.

Professional Negligence, Liability, and Indemnity

By Professor M. P. Furmston

FACULTY OF LAW, UNIVERSITY OF BRISTOL, UK

Scope of the Paper

The purpose of this paper is to set out in outline the
civil responsibility arising from the negligence of
professional men, that is, liability to pay damages by way
of compensation. The paper does not discuss criminal
responsibility, which may arise in some cases by statute,
particularly in this context under the Health and Safety
at Work etc. Act 1974 (HSWA). The modern law has undergone
radical changes since the decision of the House of Lords in
Hedley Byrne & Co. Ltd. v. Heller & Partners Ltd., (1964)
A.C. 465. This case, actually decided in 1963, and its
successors has led the Law to perform what closely resembles
a 'U' turn. These developments can best be described by
setting out first the position as it appeared to be in 1963
and then looking at ways in which the Law has developed over
the succeeding twenty years.

The Position in 1963

In 1963 it was clear that a professional man could be
liable in Contract where he was engaged to give advice for
a fee. Of course, a good many professional men gave advice
to those with whom they had no contract because there was
no fee, the best example of this being doctors acting
under the National Health Service. For historical reasons
it was also held that Barristers were not in a contractual
relationship with their clients, whom they could not sue for
their fees. Under this principle, a salaried employee could
be liable to his employer for careless professional advice
which he gave him, though in practice it would be very unusual
for an employer to sue an employee on such facts. Much advice

would of course be given by salaried employees to customers
of the employer. Again there was no doubt that the employer
was in principle liable for careless advice given by employees,
and indeed in some cases he might be liable even though the
advice was not careless on the part of the employee, for
example where a solicitor employed an unqualified clerk to
do work which requires more expertise than the clerk could be
expected to manage. This illustrates a general point that the
professional man may be careless in the way in which he organises
his office.

   In addition, there might be liability in <u>tort</u> for careless
conduct which led to personal injury or property damage. Up
until 1963 it was widely believed by English lawyers that there
was no liability in tort for negligent advice at least where it
did not lead to personal injury or property damage, or for
negligent acts which led to what lawyers have come to call pure
economic loss, that is, loss which does not involve personal
injury or property damage to the plaintiff. (A classic
example of such loss would be that suffered by Manchester United
Football Club, when the aeroplane carrying the team crashed at
Munich in 1957. The loss suffered by the individual players
was undoubtedly physical injury, for which they could recover
if they could show negligence, but the loss suffered by the Club
would be its financial stake in the players, who would of course
have a very substantial transfer fee value.)

   There were (and are) a number of important differences
between actions in contract and in tort. One relates to the
standard of care. In tort the standard of care is normally
that of the reasonable man (see below); in contract the
standard is sometimes one of strict liability, that is, a
liability independent of fault. So for instance a seller who
is in the business of selling goods impliedly undertakes that
the goods are of merchantable quality and reasonably fit for
the purpose for which they are bought. So a grocer who sells a
tin of salmon impliedly undertakes that it is reasonably fit
for the purpose of being eaten and is in breach of contract if
it has been negligently canned in Alaska so that it causes food
poisoning even though there were no steps which he could take

to check that this was not the case. However, this principle
does not normally apply in the context of professional
negligence since the courts have usually held that the
professional man only undertakes to use reasonable care and
skill and does not guarantee a result. So a doctor carrying
out an operation for a fee does not undertake that the
operation will be successful, but only that he will carry it
out as carefully as a reasonably competent doctor would. It
is possible also that there is a different test for the
difficult problem of remoteness of damage, that is, for
which consequences of an admitted tort or breach of contract
a defendant is liable. It appears that a defendant in a
tort action is liable for all the reasonably foreseeable
consequences of the tort, whereas a defendant in a contract
action will be liable only for consequences which at the time
of the contract are seen as not unlikely to result from the
breach. This appears to be a narrower rule. Thirdly,
different rules may appear in the case of limitation of
action, that is, the time within which an action must be
brought. Although the periods for actions are the same,
the starting point for an action in contract is the moment
when the contract is broken, whereas the starting point for
an action in tort is the moment when the tort is committed.
Since the tort of negligence is not committed until the
plaintiff suffers damage this can produce very different
results where there is a significant time gap between the
careless conduct and the subsequent suffering of the damage.
This can have important consequences where a plaintiff may have
an action both in contract and in tort (see below).

Developments since 1963
(a) Actions in Tort for Careless Advice. In Hedley Byrne
& Co. Ltd. v. Heller & Partners Ltd., the House of Lords held
that in principle there could be liability in tort for careless
advice. In that case the plaintiff was an advertising agent
who was engaged to place advertisements on I.T.V. on behalf
of a firm called Easipower. Under the practice for television
advertising, the plaintiff had to accept personal responsibility
for the substantial fees which were due and the creditworthiness
of Easipower was therefore of importance. The plaintiffs asked

their bank to make an enquiry of Easipower's bank, the
defendants. The defendants replied giving what was treated
as an encouraging reply; the plaintiffs placed the
advertisements and lost substantial sums of money when
Easipower went into liquidation. The House of Lords held
that on these facts the defendants would have been liable
had they not given the banker's reference on a form headed
"without responsibility". The House of Lords did not say
that all careless advice would give rise to liability. It
was necessary to show that the advice was given in circumstances
where it was reasonable for the advice receiver to rely on it and
where the advice giver could reasonably expect that it would be
relied on. It is important to note however that in Hedley Byrne
the defendants did not know to whom they were giving the advice;
they only knew that some customer of the National Westminster
Bank was having financial dealings with Easipower. This degree
of knowledge however was sufficient in general to place on them
a duty to be careful.

As is often the way with epoch-making decisions, the
years immediately following 1963 saw the courts treating
Hedley Byrne v Heller in a cautious manner. However, by the
1970s it had become accepted as a settled part of the law and
the stage was set for further expansion. Two interesting cases
may be mentioned in this context. One is Ross v Caumters
(1980) Ch. 297, in which it was held that a solicitor who
gave careless advice to a testator in a way which resulted
in the testator making an invalid will was liable to a
disappointed beneficiary who would have received property
under the will if the solicitor had not been careless. The
second case is Anns v Merton London Borough Council (1978)
A. C. 728, in which the House of Lords approved an earlier
line of decisions in the Court of Appeal that a Local Authority
might be liable to owners of buildings which were collapsing
if it could be shown that their inspectors had negligently
exercised or failed to exercise their powers of inspection
of the foundations of the building under the Public Health
Act 1936.

(b) <u>Actions for Economic Loss</u>. The thrust of the decisions
above was to erode a previously held distinction between
careless acts and careless advice. In the same way the
distinction between physical injury and property damage and
pure economic loss has also been significantly eroded. One
interesting case here is the decision of the Supreme Court
of Canada in <u>Rivtow Marine Ltd. v Washington Ironworks</u> (1973)
40 D.L.R (3d) 530. In this case the defendant was a
manufacturer of cranes who discovered that one of their models
of crane was defectively designed. They failed to tell owners
of this model of the defect promptly. The plaintiff was the
owner of such a crane. There could have been no doubt of the
defendant's liability if the crane had collapsed and caused
physical injury. In fact, what happened was that the plaintiffs
discovered the defect before anyone was injured but at a
particularly inconvenient time of the year so that they suffered
more than they would have done than if the defendants had told
them promptly. The Supreme Court was in principle willing to
allow an action on these facts but the members of the court
differed as to exactly what loss the plaintiffs could recover.

The House of Lords has moved into this area with its very
important decision in <u>Junior Brooks Ltd. v Veitchi Ltd</u>. (1982)
3 All E.R. 210. In this case the plaintiffs engaged a
contractor to build a warehouse for them and the defendants were
engaged as nominated sub-contractors (<u>i.e</u>. sub-contractors to
the main contractor chosen by the plaintiffs) to undertake the
construction of the floor of the warehouse. It was alleged
that the defendants had carelessly constructed the floor in
such a way that although it was not dangerous or likely to
cause an accident, it would require rebuilding much sooner
than it would have done if properly constructed. The House
of Lords held that if this could be established the plaintiffs
could maintain an action in tort against the defendants. It
should be noted that on these facts the plaintiffs would
normally have had an action in contract against the main
contractors. (It is thought that the plaintiffs were not
pursuing such an action because the main contractors had
gone into liquidation.) It is significant that the House of

Lords did not rely on any allegation that the floor was
dangerous, and the decision rests squarely on the defendant's
knowledge that if they did the work badly this was likely to
cause financial loss to the plaintiffs. It seems likely that
this decision will open the way to further extensions of
liability for careless conduct which causes purely financial
loss.

(c) Interrelation between Actions in Contract and Tort. The
developments in (a) and (b) above have greatly increased the
possibility that the plaintiff may on the same facts plausibly
claim both in contract and tort. At one time it was unclear
whether this would be permitted and there were decisions
suggesting that a claim for instance against a solicitor
would lie only in contract. However, these doubts appear to
have been laid at rest by the decision in Midland Bank Trust
Co. Ltd. v. Hett Stubbs & Kemp (1979) Ch. 384. In this case
a father owned two farms, one of which was farmed by one of
his sons. The father decided to grant the son an option to
purchase that farm and they together went to the family
solicitor for the option agreement to be drawn up. The
solicitor drew up an option agreement which was binding on
the parties but completely forgot that it was desirable to
register the agreement in the Land Registry so that it would
be binding on anyone to whom the father sold the farm. Some
years later the father decided to change his mind and after
taking advice from another solicitor sold the farm to his
wife for £500. It was held in other litigation that although
the wife knew of the option she was not bound by it because it
was not registered. The son brought an action against the
solicitor claiming damages for negligence. The solicitor
argued that the action against him lay in contract: that the
limitation period began on the day when he had given the bad
advice and that the writ had been issued more than six years
later, outside the limitation period. Oliver, J., rejected
this reasoning, principally on the ground that in addition to
a contract action the plaintiff had a tort action and that the
limitation period in the tort action did not start to run until
the plaintiff suffered damage, that is, when the option became
unexerciseable because of the father's sale to the mother (in
fact he also held that the contract action was not barred by
limitation, because he was able to find that the defendant had

broken the contract not only the day when he had given the bad
advice but also on every subsequent day when he failed to
register the option).

The Standard of Liability

Curiously enough, while the scope of liability has been
very significantly expanded by the developments described
in the previous section, the standard of liability has
continued to be very much the same. The classic definition
is that given by McNair, J., in Bolan v. Friern Hospital
Management Committee (1957) 2 All E. R. 118 where he said
"where you get a situation which involves the use of some
special skill or competence, then the test whether there
has been negligence or not is not the test of the man on
top of the Clapham omnibus, because he has not got this
special skill. The test is the standard of the ordinary
skilled man exercising or professing to have that special
skill. A man need not possess the highest expert skill at
the risk of being found negligent. It is well-established
law that it is sufficient that he exercises the ordinary
skill of an ordinary competent man exercising that particular
art.... there may be one or more perfectly proper standards;
and if a medical man conforms with one of those proper
standards, then he is not negligent.... a mere personal
belief that a particular technique is best is no defence
unless that belief is based on reasonable grounds.... a
doctor is not negligent if he is acting in accordance with....
a practice (accepted as proper by a responsible body of
medical men skilled in that particular art), merely because
there was a body of opinion that takes a contrary view. At
the same time, that does not mean that a medical man can
obstinately and pig-headedly carry on with some old technique
if it had been proved to be contrary to what is really
substantially the whole of informed medical opinion".

The many cases which have been concerned with professional
negligence have largely involved the working out in particular
cases of this general approach. Nevertheless, one can gather
together certain groups of problems which commonly arise.
Actual head-on attacks on the professional judgment of the
defendant are, in general, the least likely way to succeed.

Failures of management in system are a common cause of
complaint, a typical example being failure to exercise
sufficient supervision over inexperienced staff. A
professional man is also under a duty to identify problems
and warn of risks. This is a common problem for instance
with lawyers, since the legal difficulties of a situation
are often not apparent to lay clients.

Particularly relevant to the interests of chemists
is the possibility that it may be negligent not to have
carried out sufficient investigatory or research work. A
good example is to be found in the case of Vacwell Engineering
Co. Ltd. v. B.D.H. Chemicals Ltd. (1971) 1 Q.B. 88. In this
case the defendants decided to use the chemical boron
tribromide instead of boron trichloride, which they had
previously used. In fact, boron tribromide was very dangerous,
since it was liable to explode if coming into contact with even
small quantities of water. The defendants did not know this,
and the information was not to be found in a number of books
which they had looked at. On the other hand, older books
which were in the defendant's library, but which had not been
consulted, did actually record this danger. It was held that
on these facts the defendants were negligent and had not
adequately carried out their duty to research the product.
Clearly, this imposes a high standard on the defendants.

Another instructive case in regard to experimental
processes is Independent Broadcasting Authority v. B.I.C.C.
Construction Ltd (1980) 14 B.L.R. In this case the plaintiffs
were the owners of a 1250 ft television transmission mast,
which had been supplied by the first defendants and designed
by the second defendants, who were nominated sub-contractors
under the contract with the first defendants. The form of
mast design had not been used previously in this country, and
no mast of such a design and equivalent height had been built
anywhere in the world. Unfortunately, the mast broke some
two years after being erected and it was found that this was
due mainly to uneven icing of the stay-ropes, which had
caused asymmetrical tensions on the mast, and also to the
aerodynamic characteristics of the mast, which had been built
in the form of a steel cylinder, which was capable of producing

areas of turbulence at low wind speeds. The second
defendants argued that they had done as well as anyone could,
granted that the work was at the frontiers of knowledge, but
this argument was rejected. Viscount Dilhorne said "judgment
on hindsight has to be avoided.... Justice requires that we
seek to put ourselves in the position of B.I.C.C. when first
confronted by their daunting task, lacking all empirical knowledge
and adequate expert advice in dealing with the many problems
awaiting solution. But those very handicaps created a clear
duty to identify and to think through such problems, including
those of static and dynamic stresses so that the dimensions
of the 'venture into the unknown' could be adequately
assessed and the ultimate decision as to its practicability
reached. And it is no answer to say, as did one witness
regarding the conjunction of vortex shedding and ice loading,
'it wasn't obvious because it hadn't been considered'".

Insurance

It is clear that the increasing areas of liability carry
with them an increased need to take out effective insurance.
Doing so presents formidable problems. One is that liability
insurance is at the moment relatively expensive. I say
relatively expensive in that premiums appear high in relation
to the amount of cover offered. The main reason for this
appears to be that insurers find it very difficult adequately
to assess the risk. This contrasts very sharply with the
position say in relation to motor insurance where the volume
of business is so great that the amount of risk can be
assessed with a fair degree of confidence. In practice liability
insurance is nearly always offered on a limited basis for
example, up to £100,000 any one claim since unlimited liability
insurance is prohibitively expensive. This makes a realistic
evaluation of the amount of cover necessary imperative.

A second problem is that at least in some professions
cover is effected on the basis that one is covered in relation
to claims in the year of the policy and not in relation to
accidents within the year of the policy. This creates a
significant problem in relation to retired professional men
who need cover for years after they have withdrawn from practice.

This is particularly the case with professionals like
architects and engineers, the products of whose carelessness
may not collapse for a good many years until after the
original act of carelessness, which makes the operation of
the Limitation Acts a matter of crucial importance to such
professions.

Discussion

Q1.   If an accident occurs to a student in a
               university laboratory can the Professor be
               held liable?

Furmston   -   The first thing one is taught as a lawyer
               is that you don't waste time suing people
               with no money!

               The obvious course is to sue the university,
               but yes technically someone in the department
               would almost certainly be liable.

Q2.   Two of the States in the USA operate a system
               of absolute liability, in other words it is
               not enough to show that the state of knowledge
               at the time of an accident was inadequate to
               allow one to foresee the risk, one must take
               account of knowledge gained since the accident.

Furmston   -   Not so far, as far as tort actions are concerned.
               The standard tends to rise after each disaster -
               each disaster provides extra knowledge - but the
               test to be applied is what people know at the
               time of the accident.

               There is a tendency in the area of product
               liability to institute a system of strict liability
               (eg. for pharmaceuticals), when it would not be
               necessary to prove that manufacturers were
               negligent. However, I think that this would
               involve statutory changes that are several years
               away at least.

The System in the United States of America

By Professor M. Corn

SCHOOL OF HYGIENE AND PUBLIC HEALTH, JOHNS HOPKINS UNIVERSITY, BALTIMORE, MARYLAND, USA

Introduction

The administrative and technical approaches to regulation
of occupational safety and health hazards in the United
States (USA) must be discussed in the context of the periods
prior to and after adoption of the Occupational Safety and
Health Act of 1970 (OSHA of 1970). The OSHA of 1970
revolutionised the approach to this subject in the USA.
Prior to adoption of the Act occupational safety and health
regulation was almost entirely in the jurisdiction of the
states. Several states boasted excellent programs, namely
California, Illinois, Massachusetts, New York, Ohio, and
Pennsylvania. However, the majority of states did not
vigorously approach the promulgation and enforcement of
safety and health standards. During the Second World War
industrial production, rather than safety, was given the
highest priority; accident frequency and severity statistics
deteriorated. To everybody's surprise the statistics did not
improve in the post war period. A Presidential Conference
on Occupational Safety and Health was held in 1948 and a long
period of wrestling with the dimensions of the problem
ensued until the first version of the Occupational Safety
and Health Act was introduced into the Congress in 1968.
This version of the Act failed to gain approval for several
reasons, the foremost being the absence of an appeal process
for violators cited by the Secretary of Labor, who both
promulgated and enforced standards. The 1970 Act provided
for an Occupational Safety and Health Review Commission to
review contested citations.

There were selected occupational safety and health standards
in the Federal government prior to the OSHA of 1970, eg. con-

struction standards and those in the Walsh-Healey Public
Contracts Act. The Walsh-Healey Public Contracts Act applied
to contractors for the Federal Government with contracts in
excess of $10,000. The Walsh-Healey Act was never vigorously
enforced; at the time of peak enforcement there were some
60 staff compliance officers. The pre-1970 industrial health
and safety conditions in the USA are treated elsewhere,[1-3]
and will not be further discussed here.

Associations for professionals in safety and health were
founded in the twentieth century. In 1911 the American Society
of Safety Engineers (ASSE) resulted from an amalgam of other
small groups. It now numbers more than 14,000 members. The
American Industrial Hygiene Association and the American
Conference of Governmental Industrial Hygienists (2700 members)
were founded in 1939 and 1937, respectively. The American
Occupational Medical Society and the American Occupational
Nursing Society are two other highly relevant professional
associations. One must also credit individual, responsible
American industries which, during the 1930s, extended medical
responsibility beyond surgical treatment of injured employees.
Concepts which are widely accepted today were gradually
introduced, including pre-employment and periodic physical
examinations, and medical and engineering procedures to
control occupational disease. However, as late as 1970
only very large firms could boast of integrated medical,
engineering, and nursing personnel concerned with the effects
of the occupational environment on the health of the workforce.
As late as 1966-67 a survey of representative industrial firms
suggested that less than 10% of the US workforce ever came into
contact with an industrial hygienist,[4] and those who did were
affiliated with larger firms.

This brief summary of industrial conditions and professional
safety and health personnel before 1970 applied to the health
and safety of laboratory workers, as well as to that of
industrial and office employees. In fact, there has been an
underlying assumption, to my knowledge never supported by
data, that laboratory workers are probably better off than
industrial workers because they understand the difficulties

and potential hazards associated with handling potentially toxic, explosive, or flammable chemicals. Another facet of the viewpoint that laboratory workers require less health and safety oversight is that they are also more highly educated and intelligent, in general, than blue-collar workers.

The purpose of this presentation is to describe the administrative, organisational, and regulatory procedures of the Occupational Safety and Health Administration, in general, and the regulatory approach to laboratory safety and health, in particular, in the USA today.

Summary of the Occupational Safety and Health Act of 1970

The Occupational Safety and Health Act of 1970 is an ambitious, even revolutionary regulatory statute. The Introduction to the Act states that the 'Congress declares it to be its purpose and policy, through the exercise of its power to regulate commerce among the several states and with foreign nations and to provide for the general welfare, to assure so far as possible every working man and woman in the Nation safe and healthful working conditions and to preserve our human resources'.[5] The Act covers approximately sixty-eight million of the one-hundred million US workforce. Federal, State and Local government employees and some other smaller categories of the workforce are exempted from provisions of the law. Approximately five million workplaces are covered by the Act. Approximately 90% of these workplaces employ fewer than 25 persons; these account for about one-half (34 million) of the covered workforce. The other half of the workforce is employed in large establishments, which are the popular image of American industry.

The Act creates three administrative entities. The Occupational Safety and Health Administration (OSHA) is a regulatory agency located in the Department of Labor. The National Institute for Occupational Safety and Health (NIOSH) is a research organisation in the Department of Health and Human Services. The Occupational Safety and Health Review Commission (OSHRC) is an administrative adjudicative body consisting of three members appointed by and reporting to the President. Its role is to resolve conflicts between regulatees cited for violation and the regulatory agency.

Thus, the Act maintains an appeal process for the violator by
creating a Review Commission to hear appeals. In addition
to the three Commission members, the Review Commission is
staffed by administrative law judges distributed throughout
the ten Federal regions of the United States. Approximately
5% of the 247,000 citations issued last year were contested.
About 80% of those appealing receive relief.

OSHA employs approximately 2500 persons, had a $200 million
budget in the financial year (FY) 1982 and is structured as
shown in Figure 1. The agency employs about 1,400 safety and
health compliance officers in approximately 115 regional,
area, and field offices. The distribution between headquarters
and field staff is approximately 500 in the former and 2,000 in
the latter. In addition to private sector compliance activities,
OSHA develops safety and health standards and oversees State and
Federal programs. OSHA also operates a chemical analytical laboratory and an instrument calibration laboratory. Together they
employ about 80 staff. The budget for OSHA, of approximately
$210 millions in FY 1982, was reduced from $232 millions in
FY 1980.

Occupational Safety and Health Standards
The heart of the Act is the concept of permanent occupational
safety and health standards. A health standard in the context
of this Act goes far beyond the requirement for a Permissible
Exposure Limit (PEL) for the airborne concentration of a toxic
chemical, for example. Thus, the Act refers to a standard as
follows: 'any standard promulgated.... shall prescribe the
use of labels or other appropriate forms of warning as are
necessary to ensure that employees are apprised of all hazards
to which they are exposed, relevant symptoms and appropriate
emergency treatment, and proper conditions and precautions of
safe use or exposure. Where appropriate, such standards shall
also prescribe suitable protective equipment and control or
technological procedures to be used in connection with such
hazards and shall provide for monitoring or measuring employee
exposure at such locations and intervals, and in such manner

Figure 1

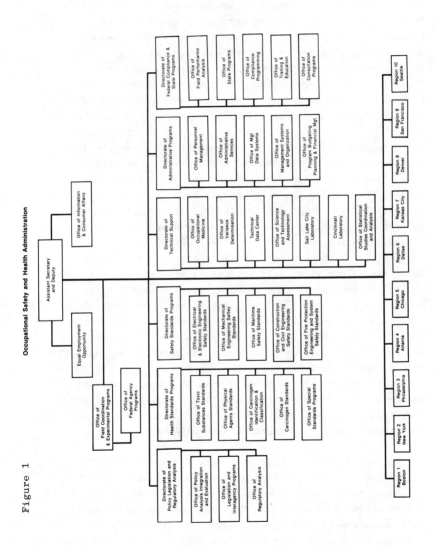

Occupational Safety and Health Administration

as may be necessary for the protection of employees. In
addition, where appropriate any such standards shall prescribe
the type and frequency of medical examinations or other tests
which should be made available by the employer or at his cost,
to employees exposed to such hazards in order to most
effectively determine whether the health of such employees
is adversely affected by such exposure'. Thus, the concept
of a standard and its ingredients, as promulgated by the
regulatory agency, were prescribed by the Congress. Each
of these requirements is very specific and can be the basis
for a violation of the standard and an associated citation.

Procedures for adoption of permanent safety and health
standards adhere to the Federal Administrative Procedures
Act. This Act requires that the agency issue a Notice of
Intended Rulemaking and a subsequent Proposal; both must
be published in the Federal Register. The Proposal should
reflect the agencies' concept of the standard to be promulgated.
A Public Hearing and a Public Comment Period follow the
publication of the Proposal. All interested parties are
permitted an opportunity to comment in writing or orally in
public and the agency must consider these comments before
promulgating a final standard. Twenty-three permanent health
standards have been issued by OSHA since its formation (Table 1).
The Toxic Substances Control Act passed in 1977 required
registration of chemicals in commercial production. Approximately 65,000 chemicals have been registered. The vast majority
of potentially hazardous chemicals in US commerce are still
unregulated. OSHA does enforce PELs for approximately 400
additional airborne chemicals, but they do not have the other
requirements which are features of a permanent standard.

During the first two years Congress permitted adoption
of previously existing Federal standards and National
Consensus Standards. Thus, many Federal standards, eg. Walsh-
Healey Public Contracts Act, National Construction Act, and
many consensus standards, eg. National Fire Protection
Association, American National Standards Institute, American
Society for Testing Materials were adopted. These standards
caused great difficulties because they were, in effect,

Table 1 - Summary of permanent health standards
promulgated since 1970[6]

Standards completed	Standards proposed, but not completed	Standards being developed
Asbestos	Arsenic [a]	Ethylene oxide
Vinyl chloride	Beryllium	Asbestos
Arsenic [a]	Sulfur dioxide	Ethylene dibromide
Benzene	Ketones	Cotton dust, non-textile sectors
Coke-oven emission	Hearing conservation (noise)	
Fourteen carcinogens	Toluene	
Lead	Ammonia	
Cotton dust	MOCA	
DBCP	Trichloroethylene	
Acrylonitrile		

[a] The arsenic standard was remanded to OSHA by the Court of Appeals for the Ninth Circuit for purposes of making a significant-risk determination consistent with the Supreme Court's benzene decision.

guidelines, and not suitable for legal enforcement. Many
have been revised to a form appropriate for regulatory
effort, eg. safety standards for walking and working
surfaces, fire protection, and machine guarding. A
significant part of the initial resistance encountered by
OSHA and its early, unprofessional reputation were based
on the large number of standards adopted during the initial
two years of agency effort.

A recent Supreme Court decision (Donovan versus American
Society of Textile Manufacturers) indicated that cost-benefit
analysis for standards promulgation is not required of OSHA
because the Congress in passing the Act did not have the
balancing of benefits to health and costs of control in mind.
However, a previous Supreme Court decision (Marshall versus
American Petroleum Institute) indicated that the agency must
provide substantial evidence that an agent causes detrimental
effects before a standard is promulgated. The Supreme Court
supported the decision of the Fifth Circuit Court to rescind
OSHA's lowering of the airborne standard for benzene from
10 p.p.m. to 1 p.p.m. These two Supreme Court decisions
have established the ground rules for promulgation of OSHA
standards.

Enforcement
The Act establishes a force of health and safety compliance
officers (CSHOs), who visit establishments to determine
whether they are in compliance with adopted standards. During
the year October 1981 through September 1982 approximately
60,000 workplaces were inspected (Table 2). Inspections can
occur because of general scheduling by the agency or as a
result of a complaint by an employee. Table 2 reveals that
in that period close to one-half of agency inspections were
at construction sites; health inspections numbered 21.1%
of total inspections. During an inspection the compliance
officer walks through the premises and may also obtain samples
of the atmosphere (health inspection). Photographs of
violations are obtained for documentation and employees are
interviewed to determine problem areas. A violation can lead
to a civil or criminal penalty; they are classified as

Table 2 - Federal OSHA compliance activity from
October 1981 to September 1982

Total inspections	61,225
New - initial inspections	21,715
Repeat - initial inspections	29,501
Follow-up	1,567
Inspections by category	
Safety inspections	43,583
Health inspections	9,200
Inspections by type	
Accident	1,879
Complaint	6,761
General Schedule	42,576
Follow-up	1,567
Inspections by industry	
Construction	29,297
Maritime	847
Manufacturing	18,013
Other	4,626
Inspection results	
Inspections with serious citations	12,864
Inspections with wilful citations	88
Inspections with repeat citations	811
Number of serious violations	22,542
Total all violations	247,717
Total penalties	$5,579.882
Employees covered by inspections	2,234,182
Complaints	
Safety received	7,067
Health received	5,159

'Other Than Serious' or 'Serious'. A Serious violation is
one having a high probability of affecting life or limb.
An Other Than Serious violation can affect the health of
the employee, but not in the manner previously described.
Compliance officers are either safety specialists or
industrial hygienists. Through cross-over training the
agency has given individuals in each of these categories
some training and understanding of the discipline of their
counterparts. The purpose of this effort was to qualify
safety specialists to refer complex situations to hygienists
and vice versa. Because of the numbers given in Table 2,
ie. 60,000 annual inspections and agency jurisdiction for
about 5 million establishments, the targeting of compliance
inspections has been a controversial aspect of OSHA
activities since its formation.

Role of the States
Section 18 of the Act permits states to assume responsibility
for health and safety regulatory programs, consistent with
the philosophy in the USA that responsibility should be
assigned to the lowest level of government that is capable
of discharging the responsibility. The law provides that
states 'be at least as effective as the Federal program'.
Federal OHSA is assigned responsibility for monitoring the
states. A state desiring to have its own occupational safety
and health program must first demonstrate that it has
established the structure of a regulatory agency, allocated
adequate positions, established standards, appeals mechanism,
etc. It must then demonstrate success at implementing the
program during a minimum period of three years, also with
monitoring by Federal OSHA. Finally, the state is given
freedom of jurisdiction. The Federal government will pay
50% of the cost of such programs during developmental and
demonstration periods. Twenty-four states have elected to
operate their own programs; several are now operating
independent of Federal OSHA. Some of the state programs
are more stringent than that of the Federal government,
eg. California, with respect of both the requirements of
standards and the vigor with which standards are enforced.

The states perform more inspections, in total, than does
Federal OSHA.

Under the Act the Federal government cannot offer
consultation. Compliance safety and health officers must
cite for violations they see after entering a facility.
However, the Federal government has been transferring funds
to every state in the Union to permit states to offer
consultation services to employers. The Federal government
contributes nine dollars to a state for every dollar the
state contributes to consultation. The consultation activities
of the states are independent of their regulatory activities,
unless a serious violation is involved. In that case, the
consultant is supposed to inform the regulatory agency in the
state of a violation if it is not abated within the time
period specified by the consultant. In general, the
consultation programs in the United States receive high
grades from employers; the regulatory programs do not. The
number of consultations performed by the states cannot be
presented because statistics are not available.

NIOSH and Research in Occupational Safety and Health

The Congress conceptualised NIOSH as the research arm of OSHA.
During recent years the budget of NIOSH has been drastically
cut, from $82 million in 1980 to $52 million in 1982 and there
are projections of further cuts. NIOSH has about 750 employees
in offices and laboratories in Atlanta, Georgia and Cincinnati,
Ohio and in ten Regional Offices. NIOSH performs research,
administers research grants and contracts, supports professional
educational efforts in occupational medicine, safety, nursing,
and industrial hygiene primarily through grants to fifteen
university-based Educational Resource Centers (ERCs), and
performs health hazard evaluations in the field. NIOSH recently
published the ten leading work-related diseases and injuries in
the USA (Table 3). It will base its program priorities and
allocate resources on the basis of the Table 3 listings.

NIOSH is perhaps best known abroad for the more than one
hundred Criteria Documents that it has published. However, its
contributions to professional education through the ERCs have

been major, adding thousands of qualified practitioners
to the field.

Table 3 - The ten leading work-related diseases and
injuries - USA 1982[8]

1.	Occupational lung diseases: asbestosis, byssinosis, silicosis, coal workers' pneumoconiosis, lung cancer, occupational asthma	6.	Disorders of reproduction: infertility, spontaneous abortion, teratogenesis
2.	Musculoskeletal injuries: disorders of the back, trunk, upper extremity, neck, lower extremity; traumatically induced Raynaud's phenomenon	7.	Neurotoxic disorders: peripheral neuropathy, toxic encephalitis, psychoses, extreme personality changes (Exposure-related)
3.	Occupational cancers (other than lung): leukemia; mesothelioma; cancers of the bladder, nose, and liver	8.	Noise-induced loss of hearing
4.	Amputations, fractures, eye loss, lacerations, and traumatic deaths	9.	Dermatologic conditions: dermatoses, burns (scaldings), chemical burns, contusions (abrasions)
5.	Cardiovascular diseases: hypertension, coronary artery disease, acute myocardial infarction	10.	Psychologic disorders: neuroses, personality disorders, alcoholism, drug dependency

Other Federal Environmental Regulations Applicable to Laboratories

One cannot examine occupational safety and health regulations
in the USA without considering the complete net of Federal
statutes adopted by the Congress during the 1960s and 1970s to
control the flow of potentially toxic chemicals in the country
(Figure 2).

Many of these laws have impact on the health and safety
of the community as they relate to the chemical laboratory.
Thus, the Toxic Substances Control Act includes a Pre-Manufacture
Notice which requires that the manufacturer provide evidence
that the chemical compound to be manufactured has been examined
for its health and safety implications, and that the potential
vendor attests to its safe marketability and usage by the
general public. The Resources Conservation and Recovery Act
has extensive provisions for disposal of hazardous waste,
including a so called 'cradle to grave' manifest system for
tracking wastes. The Clean Air Act determines the types and
quantities of emissions that can be discharged from a facility;

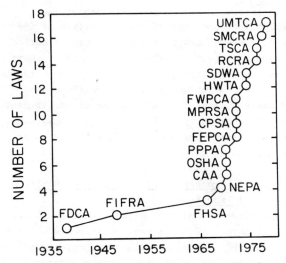

Figure 2 - <u>Acronyms for health and environmental protection laws</u>:
FDCA, Federal Drug and Cosmetics (1938); FIFRA, Federal
Insecticide, Fungicide and Rodenticide (1948, 1972, 1975);
FHSA, Federal Hazardous Substances (1966); NEPA, National
Environmental Policy (1969); PPPA, Poisonous Packaging
Prevention (1970); OSHA, Occupational Safety and Health
(1970); CAA, Clean Air (1970, 1977); FWPCA, Federal Water
Pollution Control (now Clean Water) (1972, 1977); MPRSA,
Marine Protection, Research and Sanctuaries (1972); CPSA,
Consumer Product Safety (1972); FEPCA, Federal Environmental
Pollution Control (1972); SDWA, Safe Drinking Water (1974, 1977);
HMTA, Hazardous Materials Transportation (1974); RCRA, Resource
Conservation and Recovery (1976, 1979); TSCA, Toxic Substance
Control (1977); SMCRA, Surface Mine Control and Reclamation
(1977); UMTCA, Uranium Mill Tailings Control (1978).
(From M. Corn, <u>Ann. Occ. Hyg. (London)</u>, In Press.)

it also establishes Community Ambient Air Quality Standards
and identifies Hazardous Air Pollutants. The Clean Water Act
indicates the types and quantities of emissions that can be
discharged into waterways and sewage systems adjacent to
facilities. The point of displaying these statutes is that
those concerned with laboratory health and safety must adopt
a plan for the total <u>control</u> of chemical distribution in the
facility. One cannot only consider the health and safety

impact on the laboratory worker; one must look at the
potential impact on the community of the activities of the
laboratory worker. This category of concerns can be called
'chemical control procedures'[8]. Other ingredients of chemical
controls are ordering practices, receipt and storage guidelines,
labelling, employee education and training, environmental
monitoring, and waste disposal. Distribution of educational
material to the local community to permit it to understand the
nature of potential hazards posed by the facility is being
performed to an increasing extent.

Status and Content of a Proposed OSHA Standard for Health
Safety in the Chemical Laboratory

OHSA regulates exposures to toxic chemicals under the General
Industry health standards. Some of the requirements alleged to
be inappropriate to laboratories are change rooms, shower rooms,
and special clothing changes. The carcinogen standards call for
a whole-body shower at the end of the workday. Respirators are
required, there are sign-in/sign-out requirements, medical
surveillance and records specifications, employee training,
and specific housekeeping requirements. In response to the
request for special requirements for laboratories, OSHA has
prepared (but not yet adopted) a proposed standard for laboratories
using OSHA-regulated chemicals.

    The rationalisation given for a different laboratory
standard includes the following differences between laboratories
and most industrial workplaces:

1. In the laboratory, types and classes of toxic
    chemicals are usually more numerous and they are
    used in smaller amounts.

2   Both the chemicals and the procedures used tend
    to change frequently.

3. It is frequently impossible to predict what chemicals
    or processes will be used in the near future.

4. Sufficient training to permit them to work safely
    is normally assumed to be part of the professional
    competence and responsibility of laboratory personnel.

However, OSHA cites a large number of investigations that indicate that chemists experience increased risks of cancer and other diseases. The investigations of Olin, [10-12] (N), Hoar, [13] and Searle, [14] are cited, among others. OSHA concludes that a standard for laboratories is needed, but that it must be written in such a way that decisions as to how risks of exposure are minimised are made primarily at the local level. The draft of the proposed standard (dated 10/07/82) has been circulating for comment. I will briefly review its contents.

The standard deals only with protection against toxic effects of laboratory chemicals. OSHA requirements for safety hazards are deemed to be applicable to laboratories, as well. OSHA rejects the concept of a single, comprehensive, laboratory health and safety standard. Instead, a flexible standard is developed which places primary emphasis on development and submission to OSHA for review of a Chemical Hygiene Plan appropriate to the facility. OSHA would not approve or disapprove of the plan, but OSHA does make non-mandatory recommendations concerning chemical hygiene; the implication is that they be part of the plan. Except for some special requirements for work with high-risk substances, no specifications concerning the laboratory facility itself are included in the mandatory portion of the standard, beyond the performance requirement that it must allow adequate control of exposures. The PELs of chemicals covered by OSHA standards must be met. For carcinogens, a PEL is not specified. Instead, exposures are minimised through procedural requirements. Even in these cases, the proposed standard relies primarily on prudent decisions by employers and employees rather than on detailed regulatory requirements. Requirements of standards which have been deleted as mandatory include use of a respirator when working with the substance, contiguous change and shower rooms, full-body showers at the end of the day, clean protective clothing on each entry or re-entry into the regulated area, and specific provisions concerning signs, training, emergency procedures, and maintenance.

Requirements for high-risk chemicals which have been retained from existing standards as mandatory include restriction

of work with the chemicals to limited access areas; to
conduct all such work in containment devices (hoods, glove
boxes, etc.); to decontaminate exhaust streams before
their release; to wear protective clothing, including gloves,
when working with the substances; and to follow certain
personal hygiene requirements. In addition, requirements
have been added for obtaining approval of the Chemical
Hygiene Plan by The Chemical Hygiene Officer. The latter is
designated for the facility by employer and employees by
methods unspecified by OSHA. The Plan should include, as a
minimum, provisions for:

1. Training.
2. Medical evaluation.
3. Basic laboratory chemical hygiene rules and
    procedures for handling OSHA-regulated chemicals
    in the laboratory to which the Plan applies.

In the draft proposed standard the substances for which
medical examination must be offered are shown in Table 4.

While it is always uncertain to predict OSHA policy, it
is likely that the proposed standard or one similar to it for
laboratories will be promulgated by OSHA in the short-term,
ie. 1-3 years.

Guidelines for Laboratory Health and Safety
The National Academy of Sciences has recommended procedures
for safe handling and disposal of toxic substances in laboratories.[15]
They are consensus recommendations by an expert committee
composed of academic, industrial, and government scientists.
They include both general guidelines and specific recommendations
for substances of high chronic toxicity. This document is highly
regarded.

The Department of Health and Human Services has published
'Guidelines for the Laboratory Use of Chemical Carcinogens' and
the National Institutes of Health developed and published 'NIH
Guidelines for the Laboratory Use of Chemical Carcinogens'.
These publications, although not enforceable as standards, are

Table 4 - Substances for which medical examination must be
offered if exposure or estimated exposure exceeds the level
indicated

OSHA-regulated substance	Critical exposure level
2-Acetylaminofluorene	1%***
Acrylonitrile	Action level**
4-Aminodiphenyl	0.1%***
Arsenic (inorganic)	Action level**
Asbestos	PEL**
Benzidine	0.1%***
Bis-chloromethyl ether	0.1%***
1, 2-Dibromo-3-chloropropane	PEL**
3, 3'-Dichlorobenzidine (and its salts)	1%***
4-Dimethylaminoazobenzene	1%***
Ethyleneimine	1%***
Lead	Action level
Methyl chloromethyl ether	0.1%***
α -Naphthylamine	1%***
β -Naphthylamine	0.1%***
4-Nitrobiphenyl	0.1%***
N-Nitrosodimethylamine	1%***
β -Propiolactone	1%***
Vinyl chloride	Action level*

F/N * Levels which, under the Standard for General Industry
      require initiation of medical surveillance. (Action
      level = ½ PEL).

    ** As defined by the corresponding General Industry
       Standard.

   *** Applies only to substances containing the chemical
       at or above a concentration of 1% or 0.1% as
       indicated above: examination must be offered after
       any skin contact or any usage allowing entry of any
       amount of the substance into the air inhaled by the
       employee for any period of time (such usages shall
       be considered to include work conducted in open vessels
       in a fume hood with raised sash but not work conducted
       in properly operated glove boxes or other work in which
       the atmosphere above the substance is at all times
       separated by an impermeable barrier from the atmosphere
       inhaled by employees).

also highly regarded by the scientific community and will,
hopefully, be utilised.

Conclusions

Laboratory health and safety has not received major emphasis by
OSHA, the health and safety regulatory agency in the USA. The
Occupational Safety and Health Act extends to laboratory workers,
but they are protected by General Industry Standards, designed
for different types of workplaces. A draft OSHA standard for
laboratories using OSHA-regulated chemicals exists, but remains
to be promulgated. It offers greater flexibility than do
General Industry Standards. Highly regarded guidelines for
laboratories handling highly toxic chemicals have been published
by the Department of Health and Human Services and by the
National Institutes of Health, respectively, but they are not
enforceable as standards. A legally enforceable OSHA standard
for laboratories will probably be promulgated in the next 1-3
years because several investigations indicate that laboratory
workers are exposed to higher than average risk from carcinogens
and perhaps other toxic chemicals.

References

1. M.Corn, 'Reflections on Trends in Occupational Health
    in the US: A Personal View', Ann. Occ. Hyg. 1973, 16, 251.

2. R.DeReamer, 'Modern Safety and Health Technology', J.Wiley
    & Sons, New York, 1980, pp. 3-49.

3. 'Accident Prevention Manual for Industrial Operations:
    Administration and Program', National Safety Council (US),
    Chicago, Ill'. 8th Edn., 1982, pp. 2-21.

4. V. M. Trasko, 'Resurvey of Industrial Hygiene Services
    in Industry', Am. Ind. Hyg. Assoc. J.' 1966, 27, 369

5. Public Law 91-596, Occupational Safety and Health Act of
    1970, 91st Congress, Dec. 29, 1970. Sec. 2(b).

6. 'Risk Assessment In The Federal Government: Managing the
    Process', National Academy of Sciences, Washington, D.C.'
    1983, p. 97.

7. Occupational Safety and Health Reporter, Bureau of National Affairs, Washington, D.C. 20037, January 20, 1983, pp. 711-712.

8. Morbidity and Mortality Weekly Report, Center for Disease Control, USPHS, Atlanta, Georgia, January 21, 1983, 32, (No. 2,) p. 25.

9. M Corn and P. S. J. Lees, 'The Industrial Hygiene Audit: Purposes and Implementation', Am. Ind. Hyg. Assoc. J.' 1983, 44, 135.

10. R Olin, 'Leukemia and Hodgkin's Disease Among Swedish Chemistry Graduates', Lancet, 1976, (ii), 916.

11. R Olin, 'The Hazards of a Chemical Laboratory Environment: A Study of the Mortality in Two Cohorts of Swedish Chemists', Am. Ind. Hyg. Assoc. J. 1978, 39, 557.

12. R Olin, 'The Cancer Mortality Among Swedish Chemists Graduated During Three Decades', Environ. Res., 1980, 22, 154.

13. S Hoar, 'A Retrospective Cohort Study of Mortality and Cancer: Incidence Among Chemists', J. Occup. Med,, 1981, 23, 485.

14. C. Searle, et al., 'Epidemiological Study of the Mortality of British Chemists', Brit. J. Cancer, 1978, 38.

15. 'Prudent Practices for Handling Hazardous Chemicals in Laboratories', National Academy of Sciences, Washington, D.C., 1981.

Discussion

Q1.   Are employees in organisations with less than
           25 employees usually unionised?

Corn -   The USA does not have a very unionised workforce
           - in fact the trend is towards less unionisation.
           Usually the workforce in small organisations is
           not unionised. It does seem to me that the small,
           non-unionised organisations present the major
           problem in the US.

Q2.   Can you tell us something about the practical
           problems of determining compliance with
           regulations?

Corn -   The difficulty of administering the large number
           of standards that we have depends largely on
           the professionalism of the staff. In the early
           years this was rather poor but nowadays the
           inspectors are good and are quite well respected
           by industry. If that respect isn't there the
           whole system fails. However, Compliance Officers
           are still seen as policemen rather than colleagues.
           Remember that a Compliance Officer must issue
           citations - he is allowed no discretion. Indeed
           if he does not cite for violation he is breaking
           the law. He cannot just say to the employer
           'take care of that', he must cite in writing.

Q3.   Is the number of citations generally higher in
           small organisations than in larger ones? Could
           you also give us an example of a wilful violation?

Corn -   In general it is becoming rare for large organisations
           to receive a citation for a serious violation. It
           is the small organisations that are a problem in
           this respect.

There are many types of wilful violation. A
common example is the contractor who ignores a
citation to use vacuum rather than compressed
air to clear up asbestos.

The System in the United Kingdom

By Mr J. McArdle

MANAGER OF SERVICES, SHELL UK LTD, CARRINGTON WORKS, UK

Before speaking about the system in the UK, I think it is important to tell you that I address you not as a legal expert on UK law on health, safety, and welfare at work but rather as a practitioner who has had responsibility for a considerable number of years in the oil and chemical industry for the safety and efficient operation of manufacturing plant and for essential services, such as a fire and safety department, which are so vital for running that plant.

Also, before coming to the system as it is today, we need to take a moment to look at the background to that system. As in all industrialised countries, we have in the UK seen the tremendous impact of technological change, the creation of new industries, the increased scale and complexity of industries such as oil and chemicals, and the disappearance of manufacturing plant based on obsolete technology. These developments have brought in their turn a whole host of new safety and environmental problems for a society which recognises and accepts the need for industry to work within a regulatory framework for the benefit both of employees and the public at large. However, it had also been recognised by people in government and industry that many of our regulations were designed to control specific industrial hazards in possibly obsolete plant, and were not appropriate to the second half of the twentieth century. Further, the introduction of even more regulations to control the new hazards would create an intolerable burden.

To bring an entirely fresh approach to health and safety in the workplace, in 1970 the Government appointed a Committee of Inquiry under Lord Robens with the following terms of reference:

-To review the provision made for the safety and health of persons in the course of their employment (other than transport workers while directly engaged on transport operations and who are covered by other provisions) and to consider whether any changes are needed in:

(1)   the scope or nature of the major relevant enactments, or

(2)   the nature and extent of voluntary action concerned with these matters.

and

to consider whether any further steps are required to safeguard members of the public from hazards, other than general environmental pollution, arising in connection with activities in industrial and commercial premises and construction sites, and to make recommendations.

The report of the Committee was published in 1972 and its recommendations were largely incorporated in the Health and Safety at Work etc.Act, of 1974.   This was the first stage in the process of establishing a new legislative framework.

The basic objective of the report, and of the Act of 1974, was simply to bring about a major change in the way health and safety were handled;   perhaps the most fundamental of many perceptive observations in the report was: 'The primary responsibility for doing something about the present levels of occupational accidents and disease lies with those who create the risks and those who work with them'.   The need for a change from state regulation to personal responsibility and voluntary effort on the part of managements and employees could not have been more strongly emphasised.   This change was to be brought about in general by better organisation at the workplace and action at industry level and in particular by managements and employees identifying common objectives in the field of health and safety and working together towards these objectives by exercising common judgements without resorting to traditional bargaining procedures.   In other words, the report recommended

the introduction of a form of industrial democracy in the
workplace.

However, health, safety, and welfare problems could not
be solved just by people working together more co-operatively
than in the past. A new legislative framework was required and
the need for employers, employees, and their organisations to
be consulted in the development of the new legislation was
recognised as was the need for them to work together to make it
work.

The Health and Safety at Work etc. Act creates this
framework, which is concerned with;

- Securing the health, safety, and welfare of persons
  at work;

- Protecting persons other than persons at work against
  risks to health or safety arising out of or in
  connection with the activities of persons at work;

- Controlling the keeping and use of explosive or
  highly flammable or otherwise dangerous substances,
  and generally preventing the unlawful acquisition,
  possession, and use of such substances; and

- Controlling the emission into the atmosphere of
  noxious or offensive substances from premises of
  any class prescribed for these purposes.

The Act also states the requirement to replace outdated legislation with new registered codes of practice designed to
maintain or improve standards of health, safety and welfare.

Of great importance are the general duties placed on
employers to ensure, so far as is reasonably practicable, the
health, safety, and welfare at work of all employees:

- The provision and maintenance of plant and systems
  of work that are, so far as is reasonably practicable,
  safe and without risks to health;

- Arrangements for ensuring, so far as is reasonably
  practicable, safety and absence of risks to
  health in connection with the use, handling,
  storage, and transport of articles and substances;

- The provision of such information, instruction,
  training, and supervision as is necessary to ensure,
  so far as is reasonably practicable, the health and
  safety at work of his employees;

- So far as is reasonably practicable as regards any
  place of work under the employer's control, the
  maintenance of it in a condition that is safe and
  without risks to health and the provision and
  maintenance of means of access to and egress from it
  that are safe and without such risks;

- The provision and maintenance of a working environment
  for his employees that is, so far as is reasonably
  practicable, safe, without risks to health, and
  adequate as regards facilities and arrangements for
  their welfare at work.

The act also prepares the way for the regulations governing
the appointment of safety representatives, including the duty of
every employer to consult any such representatives. The means
whereby the consultation takes place will be discussed later.

The second stage in the process and key to the establishment
of the total legislative framework was the enactment of The
Safety Representatives and Safety Committees Regulations 1977,
which came into operation in October 1978. These regulations
established the right of Trade Unions to appoint safety
representatives who could request employers to establish safety
committees. Of course, many companies already had well
established safety committees in operation before the Regulations
came into force, but the way in which these committees operated
has often changed considerably because of the new legal status of
the safety representatives.

Before moving on to examining how the new legislation
has operated in practice, it is necessary to look at some of
the other essential features contained in the Act and the
Regulations. Such features are:

-   A written management policy for health and safety
    which is available to all employees;

-   The production of a statement by management on the
    organisation of health and safety in the workplace
    and the arrangements for carrying it out;

-   Inclusion of a statement on health and safety in
    company reports (not yet enacted);

-   Establishment of a unified authority to oversee the
    operation of the legislation, i.e. the Health and
    Safety Commission, the policy making body, and the
    Health and Safety Executive, the executive body;

-   The right of safety representatives to carry out
    safety inspections and investigate accidents;

-   The provision by management of information on the
    hazards of the workplace and of the materials used
    in the workplace;

-   The detailing, in Appendix 3 of the Act, of subject
    matter for subsequent regulations covering virtually
    every facet of the safety and environmental aspects
    of the design and operation of industrial processes.

It is interesting to note that the Robens Report recognised
the possible impact on UK legislation of UK entry into the
Common Market. In particular, attention was drawn to the need
to have enabling legislation, i.e. the Health and Safety at Work
etc. Act, which would facilitate the process of harmonising
national standards with EEC standards. I wonder, however,
whether the Robens Committee, in paving the way for harmonisation

by recommending the removal of obsolete legislation, and
those who drafted the Act, in introducing potential subjects
for subsequent regulation, foresaw the flood of EEC Directives,
associated UK Regulations, and new UK Regulations with which
industry is now faced. Industry and Trades Unions do
participate in the development of the new regulations, and it
can always be argued that each new set of regulations has
justifiable objectives, but has enough time been allowed for
the new system of voluntary self-regulation at work to be
properly established and consolidated?

Many people in industry had serious doubts in 1978 when
the Safety Representatives Regulations came into force and since
then many articles have been written which, in general, have
stated quite clearly that the worst fears had not been realised.
In fact, the belief expressed in the Robens Report that employers
and employees could work together to common objectives without
conflict has been borne out in many companies. Within my own
organisation, which has a number of large refining and
petrochemical manufacturing sites and research laboratories in
the UK, the legislation is judged to have worked well, but the
ways in which health and safety matters are managed to meet the
requirements of the law, although broadly similar, are not the
same at each location. Arrangements vary according to the type of
location - manufacturing, research centre, head office, etc. -
depending to some extent on the historical development of the
site and its local culture. The concern of managements, however,
at all locations is that all employees should be involved in the
objective of raising safety standards.

The Act in its present form, i.e. as amended by the
Employment Protection Act of 1975, only confers rights on safety
representatives appointed by recognised trade unions and not on
non-unionised employees. However, the spirit of the Robens Report
was that the proposed legislation should apply to all employees
in an enterprise who should be encouraged to work together in a
voluntary and self-regulatory way. Accordingly the arrangements
for health and safety in large organisations such as Shell are
generally structured to ensure that employee representation and
involvement in the safety process are broadly similar and the

rights enjoyed by union-appointed safety representatives are
no greater than those enjoyed by non-unionised employees.

Let us now look at how safety may be managed in accordance
with the requirements of the Act in a manufacturing plant,
many of whose employees are members of their appropriate trade
unions. The actual organisation and arrangements for health
and safety are likely to be somewhat different today from what
they were in 1974, when the Act first came in, and from 1978
when the Safety Representatives Regulations became operative.
There has clearly been a period of settling down or evolution in
many organisations and those which have experienced least
difficulty are the ones in which a top management commitment to
good safety and health practice, law or no law, has existed for
a long time and has been consistently conveyed to employees.
What the law has done, however, is to replace what many
managements may previously have seen as a sole right to manage
with a requirement to work together with employees on safety and
health matters. The extent to which this change has been
accepted is probably a measure of the success which managements
are likely to have had in working to the requirements of the
legislation.

Assuming then that the principle of joint participation in
safety has been established, how can this be built on? The
minimum requirements of properly appointed safety representatives,
establishment of a Safety Committee, a statement of management
policy, and an arrangement statement, which I shall explain later,
may satisfy the law but are not in themselves any guarantee of
success in managing safety. Other aspects of the total workplace
environment have to be considered. It is implicit in the law
that managements and employees should work together in a non-
confrontational way but, if the prevailing industrial relations
climate is confrontational, it will be much more difficult to
behave differently in safety matters. Management leadership is
essential if progress is to be made and this leadership can be
exerted by demonstrating a willingness to establish and work to
joint objectives in safety even in what in other respects is a
somewhat confrontational climate. At the same time, situations
can arise where safety and industrial relations issues become
interlinked; this is a problem area in which one cannot offer
any simple advice.

Apart from the situation in the process industries now being discussed, there is an enormously varying range of tasks being performed in areas such as construction, engineering, agriculture, service, and office-based industries. The hazards to which the worker is exposed may be purely physical, a combination of physical and environmental exposure, i.e. to chemicals, gases, dusts, noise, etc., or they may be largely environmental as in laboratories. The balance of these hazards has to be taken into consideration in evolving sound safety practices, and the process industries, in which both physical and environmental hazards occur, probably offer as great a challenge as any to employers and employees alike to make the Act work effectively.

Coming back now to the workplace let us look at the individual elements which should contribute to a safety management system carrying a high level of management and employee commitment.

First, and so obvious that it should not require stating, safety is a line management responsibility under the law: others in an organisation, particularly in a safety department, may have responsibilities in the field of safety but these responsibilities are largely advisory, not executive. It is therefore logical that managements should be required to produce a written statement of their policy which, if realistic in terms of the commercial world in which companies operate but sensitive to the feelings of employees on health and safety, will be regarded by employees as a management commitment and not as a motherhood statement. Clearly managements will not make, and are not expected by the law to make, commitments to spend unlimited time and money on health and safety but they are expected to devote adequate resources to this subject to ensure that the workplace, as far as is reasonably practicable, is a safe and healthy one for employees.

The basic legal requirement is for a Safety Committee to be established if requested by employee representatives. In a large manufacturing establishment, with a considerable variety of types of work and employees represented by a number of unions,

a single Safety Committee in which everyone's interests are
catered for is likely to be too large and unlikely to be
effective.

Therefore consideration has to be given to setting up a
number of area Safety Committees, each covering common interest
groups such as manufacturing operations, maintenance, laboratory,
administration, etc. In each of these committees, employees
and management at all levels are represented and their interests
are in turn represented by delegates to the main Safety Committee,
whose size should be contained to 12-18 members. The setting up
of this type of structure cannot be brought about by a
management decision alone; for the system to be effective, there
must be full discussion and agreement with employee representatives.

Having established a representational structure, which
hopefully will not result in unacceptable incursions into
management and employee time, further consultation should follow
on the arrangements for meeting legislative requirements - the
so-called arrangement statement. This statement should include
the company policy, a definition of the duties and responsibilities
of people at all levels in the organisation, including safety
representatives, descriptions of the committee and employee
consultation structure, training arrangements, general information
on safety and health at work and workplace hazards, and medical
arrangements. There can be other types of information,
appropriate to individual industries or workplaces, which can be
included in the arrangement statement; of these I suggest that
the most useful are guidelines for resolving problems, with
emphasis on a co-operative rather than a procedural approach,
and guidelines on disclosure of information. There are constraints
which must be applied to disclosure of information by managements
to employees, whether or not they are trade union representatives,
and these constraints need to be recognised.

Of course, general arrangements for safety and health at
work, although broadly acceptable to employees and trade unions,
are not enough in themselves to have an impact on safety
attitudes and performance. From a management point of view,
there has to be a professional safety department whose role in

safety management is primarily advisory and recognised by
everyone to be impartial. Inevitably, situations arise in
which the boundary between safety and industrial relations
issues is diffuse; it is not the role of the safety adviser
to 'take sides' but to encourage resolution of such problems
within the line management system and according to the spirit
of the law.

Training is another prime area requiring management
attention if the needs of the business and the requirements of
the law are to be met satisfactorily. While a company may
consider that its in-house training facilities can cope
adequately with the safety training requirements of employees,
including safety representatives, demands will probably be made
for external trade union organised training courses and I believe
that a reasonable employer should seek to strike a balance between
in-house and external training. There are no guidelines, however,
to indicate what might be regarded as a reasonable amount of time
to be spent on safety, or any other form of training, and every
management has to decide both what it needs to do in this area
and what it can afford. What no management can afford, however,
is to pay less attention to the training of foremen, supervisors,
and managers than to the training of safety representatives.

The law requires that there should be a statement of company
policy and an arrangements statement which need not necessarily
carry the level of employee commitment which both management
and the law are seeking, while, on the other hand, there is
likely to be greater employee commitment to the arrangements
statement if there has been employee participation in its
production. To increase the potential effectiveness of the
policy and arrangements statements, a process of objective setting
is recommended in which employees and managers jointly set
practical, realisable objectives which, if achieved, should
contribute towards improving the safety performance of the
enterprise.

To complete the scenario in which safety is managed, we
need to look at the role of the safety representative in more
detail. The legally established functions of the safety

representative are clearly stated in the regulations and
may be summarised as representing the interests of employees
to employers on matters of health, safety, and welfare at the
workplace. In performing these functions, the safety
representative has rights without any real responsibilities in
law. Perhaps it was this aspect of the new safety representative
regulations which initially gave rise to so much concern among
managers who felt that the seeds of confrontation had been sown
and that the self-regulatory approach envisaged by Robens was
being undermined. In practice, it has not worked out this way
because, for the most part, safety representatives who have
been faced with and accepted the real responsibility of their
position have found that there is no sensible alternative to
working co-operatively with managers to improve the safety
situation in the workplace. For their part, more senior managers
have found that the system cannot work effectively unless the
junior managers who have the most frequent contact with safety
representatives have a good working relationship with them.

As I have already mentioned, the Health and Safety at Work
etc. Act established what are described in the legislation as
"two bodies corporate" called the Health and Safety Commission
(HSC) and the Health and Safety Executive (HSE) whose functions
shall be performed on behalf of the Crown.

The HSC consists of employer, employee, and local authority
representatives, with a chairman appointed by the Secretary of
State for Employment. This body can be regarded as the national
health, safety, and welfare policy-forming group which
determines the areas in which new legislation will be developed
and is advised for this purpose by a number of advisory committees,
e.g. the Advisory Committee on Major Hazards.

The Health and Safety Executive is a professionally staffed
body which operates on behalf of the Health and Safety Commission
in ensuring, via its Inspectorates, that the requirements of the
laws relating to health, safety, and welfare at work are met.
Individual sections within the HSE deal with factories, air
pollution, major hazards, and medical matters. These activities
are supported by laboratory and research facilities which also
act collaboratively with industry and other bodies.

In exercising its role under the Health and Safety at
Work etc. Act, the HSE is required to ensure that employers
adhere both to the Act itself and to various health and safety
regulations and approved codes of practice which are issued by
the Secretary of State.   These codes of practice have an
extremely important part to play in safety at the workplace.
Because of the enactment of European Community Directives,
we are, of course, finding that UK regulations are being
introduced which must at least reflect the requirements of the
EEC Directives but can exceed them.   These are in addition to or
may replace purely domestic regulations.

The value of sound HSE codes of practice cannot be over-
emphasised, given that they are reasonably practicable in
application, but individual industries should also be encouraged
to develop their own internal codes of practice and employee
involvement in this process should be encouraged in order to
generate maximum all-round commitment.

If we look now at the powers held by the Factory
Inspectorate in monitoring a workplace operation for conforming
with statutory provisions there are two specific actions an
Inspector can take:

- He can issue an improvement notice requiring an
  employer to remedy a contravention within a
  specified period of time;

  or

- In more serious cases where there is a risk of
  serious personal injury, he can issue a prohibition
  notice.

The latter notice requires that the activity contravening the
relevant provision or code of practice shall not be carried on
until the necessary remedial action has been taken.   When either
improvement or prohibition notices are issued, the inspector
must give detailed reasons;   in both cases, appeal facilities
are available to the person served with the notice.

Obviously no employer would wish to be served with an improvement or prohibition notice and in an ideal world where all employers and employees conform to regulations and codes of practice this would not happen and the Inspectorate would be less busy than they are today. What then can be done to avoid the possibility of being served with a notice? I think the answer here lies not just in conforming to the letter of the law but more, in the 1980s, in the extent to which people act in the spirit of the Robens Committee Report and of the Health and Safety at Work etc. Act. Of course we must have Safety Committees, safety representatives, sound systems of safety management, management policies, objectives, good working practices, good training for employees and their representatives, adequate information for employees, and acceptable codes of practice. Above all we must have the voluntary self-regulation (now being undermined by too much regulation?) which Robens sought and this has to operate against a background of positive attitudes to safety and acceptance of the need for a disciplined approach in safety matters. I believe that enterprises which encourage these things need not worry about any enforcement activities on the part of the Factory Inspectorate. Rather, they will be rewarded with good safety performance and, as a consequence, reduced attention from the Inspectorate.

The final point on which I wish to touch is that of health and safety in the laboratory which is, after all, the subject of this symposium. While the interpretation of the law and the systems I have described have related largely to manufacturing-type operations, it is nevertheless true to say that a large laboratory, while having health and safety problems unique to its own operation, must still be run under the same legal framework and must therefore be organised, from a health and safety point of view, along similar lines. As I have already mentioned, the fact that a laboratory staff may not be strongly unionised should not mean that non-unionised people should have rights inferior to those conferred by law on union members. However, it is important in this context to recognise that research laboratories, which were not previously covered by the Factories Act and the Offices, Shops and Railway Premises Act, and their related regulations are, since the Health and Safety at Work etc. Act

was introduced, now covered by these Acts and Regulations.
Therefore, it is possible that in some organisations the new
law resulted in substantial changes in laboratory safety
management and that the problems encountered have been greater
than in those which already had such things as safety committees
and employee consultation procedures. Nevertheless, the effect
of the Act in all laboratories has been to make people generally
much more aware of their individual safety responsibilities.

Particular problems which have to be addressed in
laboratories are often in the field of product safety and use
of specialised equipment. Large numbers of chemical materials
are likely to be in use or stored in relatively small quantities.
Emphasis has to be placed, therefore, on the safe handling of
these materials and on their toxicological properties. It must
never be assumed that even the most highly qualified graduate
knows what to do: he has to be educated in these matters just
as his less skilled colleagues must be.

In conclusion, I would say that we have a legal framework
for health, safety, and welfare within which we in the UK can
work effectively. Although I would like to see some changes I
think the system should continue to work well in future and I
hope that the Robens objective of replacing regulation by
self-regulation is one that our legislators will keep in mind.

Discussion

Q1   The Robens Report recommended that companies
          should include health and safety statistics in
          their annual reports. Do you think there is
          anything to be gained by this?

McArdle - I'd like to put in the other way around. If a
          company is acting responsibly in this area it
          should have nothing to fear and should certainly
          publish such data.

The System in the Federal Republic of Germany

By Dr F. Heske

PROFESSIONAL ASSOCIATION OF THE CHEMICAL INDUSTRY, HAMBURG, FEDERAL REPUBLIC OF GERMANY

The basic law of the Federal Republic of Germany places the state under the obligation of caring for the bodily freedom of its citizens from injury, and of helping and supporting all those in need of protection. This contains the obligation of so organising the working life that employees can work free from accident and under conditions befitting human beings.

In practice the state fulfils these obligations through a system of institutions and authorities. The most important institution for labour protection is the governmental industrial supervision and for accident prevention this is the accident insurance carrier.

Governmental Industrial Supervision. Governmental industrial supervision came into being with the enactment of a Prussian law, which laid down that factory inspectors are to be appointed as 'organs of the government' for keeping a watch on the observance of this statutory regulation. In 1878, in the supplement to the Industrial Code, the introduction of factory inspectors in all German Federal States was made obligatory. Nowadays in addition to labour protection the governmental industrial supervision in most Lands of the Federal Republic of Germany undertakes environmental protection.

The central regulation for the activity of the governmental industrial supervision is entitled 'Industrial Code'. According to this Code the officers of the governmental industrial supervision have all the official powers of the local police authorities, in particular the right of inspecting and testing factory installations at all times.

In addition to the industrial supervision officers, factory doctors are also to be considered as special officials within the meaning of section 139b of the Industrial Code. In most of the Federal Lands these are not integrated in the governmental industrial supervision but are brought together in special administrative offices. Regulation of competence, organisation, and activity of the governmental industrial supervision falls within the competence of the Federal Lands.

The governmental industrial supervision offices are authorities of the individual Federal Lands. Any particular state industrial supervision office is competent for a certain district within one Federal Land. All factories established in this district are supervised independently of the branch of industry to which they belong.

Accident Insurance Carriers. The system of labour protection in factories in the Federal Republic of Germany runs on two lines. This dualism can be recognised in the first place in matters concerning the regulations and secondly in the supervision of the labour protection measures to be carried out by the factories. Here the legal aspect contained in the dual system is interesting. Beside the governmental regulations there exist other statutory regulations in the field of labour protection in the Federal Republic of Germany issued by the accident insurance carriers.

In other countries accident insurance companies also issue regulations which prescribe measures to be taken for the prevention of accidents at work and which must be observed by the employer as one of the contracting parties. However, as contract law these regulations are to be classified as part of civil law. In the German labour protection law this is not the case. Here the regulations for the prevention of accidents are issued by the statutory accident insurance carriers and therefore as statute law fall within the sphere of public law. From this it follows that the employer, who is compelled to become a member of a definite accident insurance carrier, can, in case of non-observance of the regulations for the prevention of accidents, be compelled through the means of enforcement, which are also available to the governmental industrial supervision, to observe these regulations.

The statutory accident insurance carriers in the Federal
Republic are not insurances based on private law, but are
corporations of public law. As such they are indirect
Federal authorities whose power of legislation has by operation
of law, namely the German National Insurance Code, been
conferred on their members. The Industrial Injuries Insurance
Institutes undertake sovereign tasks and are subject to the
supervision of the Federal Ministry of Labour and of the
Federal Insurance Office.

This kind of legal position for accident insurance in the
system of labour protection does not exist in other countries.
It is the result of a specifically German development. The
Industrial Injuries Insurance Institutes were created by the
Accident Insurance Law of 6th July 1884 as self-administering
corporations for combating accidents and for accident insurance.
During the following period the Industrial Injuries Insurance
Institutes issued a large number of regulations for the prevention
of accidents, the observance of which is watched over by
technical supervisory services and factory inspectors established
for this purpose.

The statutory accident insurance is part of the social
security insurance. The state has transferred the implementation
and administration to the accident insurance carriers, who carry
out their tasks on their sole responsibility under governmental
supervision.

Today there exists the following structure, which has
developed historically, of the carriers of statutory accident
insurances, who are classified into a total of 92 individual
carriers:
      34 Industrial Injuries Insurance Institutes
       1 The Mariners' Association
      19 Agricultural Injuries Insurance Institutes
      13 Communal Injuries Accident Associations
       6 Fire Brigade Accident Insurance Funds
       6 Communities with Executive Authorities
      11 Federal Lands with Executive Authorities
         the Federal Government with 4 Executive Authorities
         the Federal Institution for Labour

The Industrial Injuries Insurance Institutes are sub-divided
according to specialised fields. They are competent for factories

in a definite industrial branch independent of the Federal
Land in which they are situated.

At present there are in the industrial field 34 Industrial
Injuries Insurance Institutes. They are competent for the
following branches of industry: metal, mining, stone- and
earth-working, gas and water, chemistry, wood and cut materials,
printing and paper, textiles and leather, foodstuffs and
stimulants, building, commerce, and services, as well as traffic
and health services.

The Governmental Industrial Supervision and the Industrial
Injuries Insurance Institutes employ about 2,500 and 1,000
officers, respectively.

Labour Protection Law and Accident Prevention Law
The organisation of working life for safety is promoted in the
Federal Republic of Germany by way of two independent domains
of the law. These are the result of historical development
and exist side-by-side with equal stature. They are:

- The Labour Protection Law as part of the Labour
  law;

  and

- The Accident Prevention Law as part of the Social
  Security Law.

In the Labour Protection Law the state acts directly by
issuing legal regulations as well as through the supervisory
activity of the governmental industrial supervision.

In the Accident Prevention Law the state is itself partly
concerned with accident prevention, but has also in part
transferred sovereign tasks to accident insurance carriers, as,
for example, in the issuance of regulations for accident
prevention and their supervision by technical supervisory
services.

The requirements of the Labour Protection Law, which is

to be observed by employers and in some cases also by employees,
are to be found in laws and ordinances enacted by the various
Federal Government authorities and the authorities of the
Federal Lands.

In the Accident Prevention Law, regulations are issued
on accident prevention and first aid in the legal form of
regulations for accident prevention and administrative
regulations. Regulations for accident prevention are issued
by the self-administering corporations of the accident
insurance.

The basic provision of the Labour Protection Law is almost
the same in its content as the basic provisions of the
Accident Prevention Law:

Section 20a of the Industrial Code - No dangers for the
employees may arise from work rooms, works equipment, machines,
and appliances or from the factory;

Section 546 paragraph 1 and Section 708, of the German
National Insurance Code - Accidents at work are to be
prevented by all suitable means.

These two provisions are the foundations for subsequent
regulations. Observance of the regulations of the Labour
Protection Law and of the Accident Prevention Law is mentioned
by supervisory services through supervising officers.

During recent years the supervising officers have noted
an increase, which is sometimes considerable, in the demand
for advice. This increase is to the detriment of the supervisory work. This increase is partially due to constantly
increasing technical complexity, to the rising number of rules
and regulations, and not least to enquiries for information
on the part of the factories. A further increase in the
advisory service is expected for the future.

Rules and Regulations
To achieve the aims of labour protection, rules and regulations
are necessary. They are the most important source of knowledge

for the employers, the employees, the employees' councils,
the factory doctors, the testing stations, and the supervisory
services. In the Federal Republic of Germany there are two
kinds of regulations for protection at work and for the
protection of the health of employees in industrial workshops.
These are:

- The governmental regulations, namely laws and
   ordinances;

   and

- The regulations for accident prevention of the
   Industrial Injuries Insurance Institutes.

In the governmental regulations fields of activity are
regulated that extend over more than one branch. In general,
technical safety requirements for particular manufacturing
and processing machines or particular plant for manufacturing
and processing as well as the rules of behaviour pertaining
thereto are not regulated in governmental rules.

In the regulations for the prevention of accidents fields
of activity relating to specific industrial branches are
regulated, especially technical safety requirements on particular
manufacturing and processing machines as well as on particular
plants for manufacturing and processing and also the regulations
for behaviour pertaining thereto.

Examples of Governmental regulations include the steam
boiler ordinance, elevator ordinance, acetylene ordinance,
pressure gas ordinance, ordinance on combustible liquids,
ordinance on dangerous working media, law on explosives,
ordinance on workplaces, young workers' protection law,
mothers' protection law, and home work law. The state covers
the fields referred to in these regulations with laws and
ordinances. In such cases these are always fields that extend
over more than one branch.

Examples of regulations of the Industrial Injuries
Insurance Institutes include the following subject matters:

machines for the clothing industry, ceramics industry, sugar
industry, chemical industry, foodstuffs industry, metal-working
machines, wood-working machines, printing machines, industrial
trucks, lifting conveyors, trackways, excavators, pile drivers,
refrigeration plants; paint spraying plants, and steelworks.
According to their content the regulations for the prevention
of accidents do not extend over more than one branch, even
though there is a series of regulations for the prevention of
accidents that are important in more than one branch.

It can be seen from the lists of examples that the sets
of regulations mutually supplement each other. As a rule it
is to be assumed that one branch will be dealt with in only
one set of regulations. However, neither the governmental
ordinances nor the general administrative regulations nor
the regulations for the prevention of accidents together with
their instructions for implementation can always regulate the
actual practical technical safety measures completely in all
necessary details. In order to integrate these technical
safety and hygiene determinations into the Labour Protection
Law as fully as possible, references are made in the governmental
ordinances and general administrative regulations, as well as in
the regulations for the prevention of accidents and instructions
for implementation, to such Technical Rules, German Industry
Standards, provisions of the Association of German Electro-
technicians, and Guide Lines, as these contain the generally
recognised technical safety, factory medical, and hygiene rules.
The consequence of this is that these rules are to be just as
strictly observed by those having the duty to fulfil them as
the statutory regulations, unless the same or a higher degree
of safety can be ensured in some other way.

Demands, Administrative Acts, and Procedure in Labour
Protection and Accident Prevention
The supervisory services of the state and of the Industrial
Injuries Insurance Institutes have the task of inspecting
factories and thereby of keeping a watch on the observance
of the rules and regulations by the employer. Should it be
found that rules and regulations are not being observed the
supervisory services have to take steps by means of improvement

letters or directives to ensure that a condition complying
with the rules and regulations is established. In particularly
severe cases the supervising authorities can close down
machines and plants.

In the final analysis the aim of the work of supervision
is to keep a watch on the observance of the standards of labour
protection and, if departures from the standards have been
ascertained, to make the necessary demands for the establishment
of the desired condition. For this purpose the supervisory
services have a variety of methods at their disposal. Both the
governmental industrial supervisory officers and the factory
inspectors primarily use the method of improvement letters.
No obligatory method is prescribed on how to react in individual
cases towards non-observance of labour protection standards.
Within a certain range this is left to the individual officer
or it arises out of the mostly informal instructions of the
industrial supervision office or of the factory inspectorate
service, as the case may be. The carrying out of directives
is as a rule notified with a written announcement of fulfilment
on the part of the factories. Administrative compulsion and
financial penalties are likewise used as methods of enforcing
demands.

In general it must be stated that the structure and
effectiveness of legal facilities for the enforcement of
demands in labour protection represents no central theme
for the supervising officers. On the basis of experience
up to now it is believed that more can be achieved through
information, motivation, and advice than by the application
of coercive means. The enforcement of demands by imposing
sactions takes place therefore only in case of extreme
necessity.

Accident Prevention in Laboratories
The prevention of accidents in laboratories is regulated by
the 'General Regulations' of the Regulations for the Prevention
of Accidents of the Central Association of Industrial Injuries
Insurance Institutes. These regulations are supplemented by
the 'Guide Lines for Laboratories'. They apply to laboratories
in which dangerous materials are handled or on which work is

carried out for their analysis, preparation or mode of use
or application according to chemical or chemicophysical
methods, irrespective of whether the laboratories are independent
undertakings or parts thereof.

I do not wish to go into the details of these most important
'Guide Lines for Laboratories', but in the following I shall at
least acquaint you with the table of contents. I shall deal
in greater detail with those sections that are concerned in a
special manner with questions of health and safety in
laboratories.

Guide Lines for Laboratories

Table of Contents:

1. Range of application
2. General requirements
3. Building and equipment
    3.1 Doors, flooring, ventilation
    3.2 Fume cupboards
    3.3 Work tables and sink units for solvents
    3.4 Piping and fittings
    3.5 Emergency shower bath
    3.6 Electric installations
    3.7 Pressure containers
    3.8 Refrigerators and Dewar vessels
4. Operation of laboratory
    4.1 Instructions for operation
    4.2 Special instructions
    4.3 Glass appliances
    4.4 Heat protection and hot baths
    4.5 Hoses and fittings
    4.6 Packings and stoppers
    4.7 Setting up of apparatuses
    4.8 Keeping of stocks and safe keeping of chemicals
5. Conduct of persons employed
    5.1 Information
    5.2 Orderliness, cleanliness, and attendance
    5.3 Filling and transport of dangerous materials
    5.4 Cleaning
    5.5 Safety equipment
    5.6 Handling stoppers
    5.7 Food and drink
    5.8 Smoking
6. Clothing and footwear
    6.1 Clothing
    6.2 Footwear

7. Dangerous work
   7.1 Bomb tubes
   7.2 Pressure gas bottles and fittings
   7.3 Filling of gases in liquid condition
   7.4 Working with pressure gases
   7.5 Working with vacuum
   7.6 Working with combustible liquids
   7.7 Working with danger of explosion
   7.8 Handling dangerously explosive materials
   7.9 Drying in warming cabinets
   7.10 Deep freezing
   7.11 Working with materials dangerous to health
   7.12 Dangerous rubbish and waste

8. Fire protection
   8.1 Fire-fighting equipment
   8.2 Conduct in case of fire

9. Personal protective equipment
   9.1 Eye protection devices
   9.2 Protective gloves
   9.3 Respirators
   9.4 Protective clothing

10. First aid
   10.1 Steps to be taken and material

11. Instruction

12. Testing
   12.1 Gas taps and gas pipes
   12.2 Emergency shower baths
   12.3 Chemicals delivered

In the chapter 'Building and equipment' it is stated that laboratories must have adequate technical ventilating equipment effective at all times. The room ventilation can thereby go via the fume cupboards. For 'bench top' fume cupboards at least 400 m^3 per hour must be drawn off by the fume cupboard, for deep level fume cupboards at least 600 m^3 per hour, and for passable fume cupboards at least 700 m^3 per hour per metre of frontage length. The quantity of air drawn off must be replaceable with outside air, heated if necessary. The room ventilation must be so designed that the full capacity of the fume cupboard is maintained. Recirculated air may be used for the room ventilation only when no threatening increase in the content of dangerous substances or gases can occur.

For the windows of fume cupboards safety glass, preferably laminated glass, is to be used. Fume cupboard ceilings must be so designed that in case of deflagrations or explosions they can function as pressure-relief devices. Vertically sliding fume cupboard windows, especially front sliding panels,

must be secured against falling down. When the front sliding
panels are closed an air gap of 3-5 cm must remain at their
lower edge.

The fume cupboard must have service openings, to enable
work to be carried out inside the fume cupboard with the
front sliding panel or hinged window closed. This can be
effected, for example, by means of horizontally sliding
penels or a pendulum flap. For the setting up and operation
of tall apparatus service openings of this kind are also
recommended in the upper front window.

A device must be fitted on fume cupboards to indicate that
the extraction is functioning effectively. Examples of such
devices are a woollen thread or a small wind fan-wheel.

Places firmly installed inside the fume cupboard for
taking liquids or gases must be capable of being operated
from the front.

All work by which gases, vapours or suspended matter
that are very poisonous, poisonous, injurious to health,
corrosive, irritant, carcinogenic, mutagenic, or in any
other way injurious to human beings can arise may be carried
out only in fume cupboards. With work of this kind the front
sliding panels are to be kept closed. Such work can be carried
out outside the fume cupboard only when suitable measures have
been taken to ensure that the persons employed are not exposed
to the effect of the substances used. For example, such
measures might include the use of gas-tight and liquid-tight
apparatus, cooling traps, and effective extraction.

Work with anhydrous hydrogen fluoride and hydrofluoric
acid may be undertaken only in a fume cupboard. In addition
to protective goggles, a protective shield, long protective
gloves, and, if necessary, a full mask and additional suitable
protective clothing are to be worn.

The chapter 'Conduct of the persons employed' makes up
the most extensive part of the 'Guide Lines for Laboratories'.

Here, for example, it is laid down that food and drink must
not be prepared or kept in chemical or laboratory vessels,
such as beaker glasses, or put down in the work place. No
vessels may be used for chemicals that are normally intended
for the acceptance of food or drink.

There must be no eating or drinking at laboratory tables
or fume cupboards.

No foodstuffs, including drinks, may be brought into
laboratories that are specially dangerous to health, such
as medical and radio-chemical laboratories and laboratories
for the preparation of crop-protection chemicals.

Food and drink must not be kept together with chemicals.

The warming up of food and drink in laboratory warming
equipment, for example in warming cabinets, is not allowed.
Only refrigerators intended and marked for the purpose may
be used for the cooling of foodstuffs and drinks.

No smoking is allowed in any laboratory.

A special section is concerned with dangerous rubbish
and waste, to which some importance is attached against the
background of a healthy and safe work place. Thus, the head
of the laboratory has to take care that dangerous rubbish is
so collected and transported as completely to prevent any
danger to the persons employed, especially to the laboratory
personnel and to those concerned with the disposal of waste.

With a strict observance of all the requirements of the
'Guide Lines for Laboratories' it can be expected that safe
working in the laboratory and a healthy atmosphere at the work
place will also be possible in the future.

The 'Guide Lines for Laboratories' can be obtained in the
German language from: Carl Heymanns Verlag, Gereonstrasse 18-32,
5000 Köln 1.

Discussion

Q1.   You said that your Factory Inspectors were
        given the powers of the police. Does this
        include the power of arrest?

Heske - No. But it does include the power to close
        plants and suggest fines.

Q2.   Is there any conflict between the two agencies
        when dealing with similar problems?

Heske - No, there is not really any conflict. Generally
        at the inspector level we work as close
        colleagues. However at the political level
        there is some conflict perhaps.

Q3.   What mechanism exists within a plant to devise
        ways of operating in new situations in advance
        of regulations?

Heske - We operate guidelines which are 'frames' for
        safety allowing a certain amount of flexibility.
        The guidelines for laboratories, however, will
        not be turned into regulations.

Hazards of Handling Chemicals

By Mr. L. Bretherick

CONSULTANT (FORMERLY RESEARCH PROJECT LEADER, BRITISH PERTROLEUM
COMPANY), UK

In view of the second part of the title of this symposium,
'Where do we go from here?', I propose first to consider the
present situation concerning the hazards of handling chemicals,
and then to look forward constructively at some future
possibilities. It should be noted that my remarks apply only
to laboratories engaged on small-scale operations because
the conditions pertaining to pilot-scale and larger development
laboratories more nearly approximate to industrial-scale
operations.

Introduction

When I wrote the synopsis of this paper I was careful to
prefix the word 'hazards' with 'potential' to promote the
idea that chemicals which possess a high degree of reactivity
of one sort or another are not inevitably dangerous but may
become so if they are used in ways which are not compatible with
their particular properties. The word 'hazard' implies chance
or risk, and I think we all know that already it is possible in
most cases to reduce the risk of untoward occurrences with
chemicals in laboratories to a low and acceptable order by the
application of proper techniques, including those of management
about which my former colleague at British Petroleum,
Mr E Thompson, will have more to say later.

The potential hazards of chemicals arise either from
inherent instability or from their interaction with other
materials, which may be present as inanimate chemicals or as
more highly organised life forms. In the latter case,
untoward effects are referred to as toxicity, or burns if acute,
while in the former case the effects if severe are referred to
as fire or explosion. Because the risk of fire (which is, of

course, the rapid oxidation of a material by the oxygen
present in air) is often so much higher in most laboratories,
and the effects are so much more dramatic and immediately
damaging than toxic or other reactive effects, fire is
usually considered to be a separate and specific hazard.

So, we have the three problems of fire, reactivity, and
toxicity to contend with in our laboratory premises.
These problems may become evident before, during, or after
active chemical operations - that is in the storage, use, or
disposal phases or areas. I propose to discuss ways to control
these three problems or potential hazards separately for the
before, during, and after use phases, because the overall
conditions and especially the degrees of handling and risk
involved are often somewhat different.

Storage Hazards

One must differentiate between short-term and medium-term
storage of chemicals, and actively review the materials in both
of these categories periodically.   Short-term storage means
the day to day storage of the usually small quantities of
chemicals needed for laboratory operations which normally will
be stored there during working hours. Significant amounts of
flammable solvents - over 500 cm^3 - are often removed to
outside storage overnight. Medium-term and perhaps relatively
bulk storage is often necessary when a suite of laboratories
is serviced from a central storage area. Long-term storage
should, in my view, be actively discouraged by the review
procedure adopted, because it is often a euphemism for dumping
materials that no-one wants or can be bothered to discard.
Opportunities for long-term hoarding of chemicals are
gradually becoming economically unviable, so the problems
associated with this undesirable and perhaps antisocial practice
are likely to diminish with the passage of time. Further
examples of such an atypical relationship of problems with
elapsed time would surely be welcomed by all!

Some degree of segregation is necessary for safe
small-scale storage of chemicals in laboratories and this is

usually effected by dispersing the various bottles and
containers around the available cupboards and shelves, taking
care to separate acids from bases, oxidants from combustibles
and especially from reducing agents. Highly flammable solvents
must be kept in a fire-resisting cabinet, remembering that
the legal limit is 50 dm^3 per laboratory.

Considerably more care must be exercised over proper
segregation of the materials in a chemical storeroom, and
this topic formed one of the subjects of an American Chemical
Society symposium at Kansas City in September 1982, the
proceedings of which are being published.[1] One of the papers
dealt specifically with the related questions of fire
protection and segregation in storage of toxic and reactive
chemicals. Copies are available from this author on request
and receipt of a stamped addressed A5 size envelope.

A labelling scheme, which has recently been introduced
in the USA on an experimental basis and goes much further
than the EEC scheme, seems likely to be of great significance
for the development of good storage practice among other
desirable laboratory habits. This is the SAF-T-DATA scheme of
the J.T. Baker Chemical Company. The new labels show in
pictorial/numerical code the type and degree of fire, reactivity,
acute or chronic toxicity hazards and appropriate personal
protective equipment. The text on the label amplifies the
details and includes first aid measures. The colour of the
pictorial code panel indicates storage segregation requirements,
and a freely available full colour quarto sheet explains the
whole labelling system and its implications.[2] However, the
legal status attaching to the EEC labelling system in most of
the member states would currently preclude such expansion of
the information content of the label, though not of course
that of an accompanying leaflet or data sheet.

Although chemical storage rooms or areas are often thought
to be places where there is little activity concerned with the
use of chemicals, there may in fact be a considerable amount of

handling of chemicals connected with dispensing of part contents
of stock bottles or drums of solvents. In such cases the
provision of protective equipment (extinguishers, goggles,
gloves) and means of spill control, including wall chart
information,[3] should be at a level somewhat higher than that
appropriate to the projected activity. Information on storage
problems is also given in other references. [4, 5, 9, 10].

Hazards in Use of Chemicals

In general terms, the potential for hazards in the use of
chemicals will tend to be less in established laboratories
in which routine materials and procedures are used than in
those where new materials and new procedures are being
developed, particularly if some of the people concerned are
still developing their manipulative skills. However, it should
be possible to offset this tendency if the supervision, training
procedures, and organisational framework adopted for the
laboratory or group of laboratories, and the ready availability
to the occupants of safety information, are fully appropriate
to the perceived needs. A widely based and well motivated
Safety Committee can play a large part in all these essential
aspects, and especially that of pointing the way to relevant
information sources. A selection of sources useful for this
purpose is given in references 4-11.

Time does not permit much detail to be included on the
hazards of using chemicals, but a few examples will serve to
show that quite simple measures can be effective if sensibly
applied. Most people working in laboratories where flammable
solvents are in use will take good care to exclude ignition
sources. I well remember that a former colleague had a sign
reading 'ETHER - NO FLAMES' which he would display on the door.
But he was a good lateral thinker, (long before Dr de Bono
published his ideas on these matters) and on the back of his
sign it read 'FLAMES - NO ETHER' which could be displayed for the
inverse eventuality that he had foreseen.

Gloves are now used much more widely to protect the hands
from contact with corrosive or toxic materials than previously,
and in what I might call my formative years we developed the
habit of always washing the gloves with soap and water before

removing them, to avoid contaminating the inside of the cuff
when getting the gloves off. This precaution does not seem
to be widely practised now, perhaps because of the advent of
disposable gloves. The final example concerns the demolition
of a laboratory on the West coast of America by a graduate
chemical engineering student who wanted to heat his reaction
vessel in a molten salt bath. He put 3 lb of sodium nitrite
and 1 lb of sodium thiocyanate into his bath and started
heating it up. When melting occurred the explosion caused
structural damage to the tune of $200,000. The simple measure
which would have prevented that particular disaster was to have
looked in the Royal Society of Chemistry's book 'Hazards in
the Chemical Laboratory' (3rd Edition), where under Sodium
nitrite on page 484 it clearly states that explosion is a
likely outcome of heating this mixture of oxidising and
reducing salts. This has, in fact, been known since 1897.

This last example illustrates the rather disturbing fact
that many people working in laboratories do not seem to be
aware of or to use to proper advantage the information which
is provided and which would protect them from most of the
incidents and injuries that are too often an unwelcome feature
in the current technical literature.

Disposal Hazards

The disposal of chemical wastes, reaction residues, and
surplus chemicals has until recently received attention at
a level one or two orders of magnitude lower than that enjoyed
by the more intellectually stimulating areas of practical
chemistry. The relatively simple measures of segregating
broken glass and solid chemicals from uncontaminated laboratory
waste or of keeping flammable or insoluble liquids or
concentrated solutions of toxic metal salts out of the
laboratory drainage system may not be as widespread as we
would wish. Some technology appropriate to the safe disposal
of spilled chemicals or residues has been developed, but it
is by no means as extensive or as well known as the more
traditional areas of knowledge. There is, however, at the
moment a considerable amount of work in progress on devising

and developing methods for the disposal of particularly
hazardous materials. Some of this work was reported at a
recent meeting of the Royal Society of Chemistry Chemical
Information Group,[12] which covered both the commercial and
'do-it-yourself in the laboratory' types of activity. The
speaker from the Environmental Protection Group at Harwell
made the point that many experienced chemists appear to have
a psychological block which prevents them from accepting that
in many cases they are the person best qualified to deal with
particularly difficult by-products or residues arising from
their work - in accordance with the spirit, if not the letter,
of the Health and Safety at Work <u>etc</u>. Act 1974.

Information on disposal methods is somewhat dispersed
among various literature sources, a selection of which is
included in the reference list.[13-15] It should be noted
that reference 15 was withdrawn in 1980 and, amazingly, holders
of the book were advised to destroy their copy - anyone who
has one should treasure it as an extremely useful guide in
these matters.

Some Future Possibilities

I have indicated that there is a considerable amount of
published information available on the potential hazards
associated with the storage, use, and disposal of chemicals.
I have also mentioned my concern that there seems to be a
considerable disparity between the availability of such
information and its fully effective application to prevent
accidents in laboratories. It may now be worth spending
some time on thinking about how this disparity might be reduced.

Whereas the students of the Open University, because of the
nature of their unsupervised studies, are very fully briefed on
the potential hazards of the materials and techniques they are
using, the same is not true of many of the full-time graduates
coming into industry. Perhaps some of the detailed briefing

techniques of the Open University could be adapted and adopted
by the older seats of learning to give a closer integration of
safety aspects into the technology. Perhaps some of the
chemists in mid-career, who had completed their training before
the implications of Health and Safety at Work <u>etc</u>. Act 1974
had taken effect, would benefit from a suitable refresher
course organised by their Safety Officer or on a commercial
basis.

An existing technique, but as yet not fully developed
or widely applied, which appears to have considerable future
potential in avoiding chemical hazards, is that of using
computer programs to predict such potential hazards. There
are two programs, one American [16] and one Japanese, [17,] which
can predict with a reasonable degree of accuracy the possibility
of explosive decomposition of compounds or mixtures, or of
ignition on contact of two or more compounds, respectively.
This prediction is made solely on the basis of chemical
structure. I also recall seeing on one of the stands at the
Kansas City meeting in 1982 information to indicate that a
similar system to predict toxic hazards is also in course of
development. A further application of the ubiquitous micro-
computer might be to provide a chemical hazards package which
would provide details of hazards and preventive techniques
as well as storage and disposal procedures relevant to each
item in a particular laboratory or storeroom. This might
possibly be coupled with the ordering procedure so that the
information arrived before the new chemicals. Perhaps this
is something that the Royal Society of Chemistry Information
Services, which already produce an excellent monthly Laboratory
Hazards Bulletin, might like to consider adding to their
repertoire in due course.

When such systems are fully refined, and when practising
chemists and students have ready access to them and to the
numerous other data bases which give hazard-related data on

chemical materials, perhaps the more ideal situation where
the real potential for hazards in using chemicals will be
grasped and appreciated before starting to use them will be
more generally attained.

References

1. 'Safe Storage of Laboratory Chemicals', ed. D. A. Pipitone
    New York, John Wiley, 1984 (in press).

2. SAF-T-DATA Labeling System, J. T. Baker Chemical Co.,
    Phillipsburg, New Jersey 08865, USA.

3. 'Dealing with Spillages of Hazardous Chemicals' Poole,
    BDH Chemicals Ltd., 4th Edn. 1981 (370+ chemicals tabulated).

4. 'Handling Chemicals Safely 1980', Dutch Association of
    Safety Experts, Amsterdam, 2nd Edn., 1980 (800 chemicals
    tabulated, available from the Chemical Industries
    Association, covers storage, use, toxicity, and spill
    disposal).

5. 'Hazards in the Chemical Laboratory', ed. L. Bretherick,
    London, Royal Society of Chemistry', 3rd Edn., 1981 (covers
    fire, toxicity, reactivity, and spill disposal for 480+
    chemicals).

6. 'Patty's Industrial Hygiene and Toxicology, Vol 2,
    Toxicology', ed. G. D. and F. E. Clayton, New York, Wiley,
    3rd Edn., 1981 (individual chapters devoted to groups of
    related chemicals).

7. 'Registry of Toxic Effects of Chemical Substances' (RTECS),
    Washington DC, NIOSH, (revised annually; 1982 Edition
    has all published information in highly condensed form
    on 52,000+ compounds).

8. 'Dictionary of Organic Compounds', ed. J. Buckingham, London,
    Chapman & Hall, 5th Edn, 1982 (brief indication of hazards
    and toxicity for around 7,000 of 46,000 entries).

9. 'Prudent Practices for Handling Chemicals in Laboratories',
    Washington DC, National Research Council, 1981 (general
    text).

10. M. E. Green and A. Turk, 'Safety in Working with
    Chemicals', New York, Macmillan, 1975.

11. L. Bretherick, 'Handbook of Reactive Chemical Hazards',
    London, Butterworths, 2nd Edn., 1979 (updated Edition in
    preparation will cover some 10,000 compounds and mixtures).

12. 'The Disposal of Hazardous Wastes from Laboratories',
    London, Chemical Information Group of RSC, 1983. *

13. M. A. Armour, L. M. Browne, and G. L. Weir, 'Hazardous
    Chemicals: Information and Disposal Guide', Edmonton,
    University of Alberta, Canada T6G 2G2, 1982 (200 chemicals).

---

*   Available from Mr. H. Baxter, 59 Kenwood Gardens, Gants
    Hill, Ilford, Essex IG2 6YG (£6, cheques only with order,
    payable to Chemical Information Group).

Any further contributions to ref.[11] 'Handbook of Reactive
Chemical Hazards', should be sent to L. Bretherick,
17 Walton Drive, Ascot, Berks SL5 7PG. UK.

14.   'Catalogue Handbook of Fine Chemicals', Gillingham,
      Aldrich Chemical Co. (annual issues;   1983-4 Edition has
      brief indication of flammability, toxicity, and disposal
      method for each of 12,000+ items listed).

15.   'Laboratory Waste Disposal Manual', Washington DC, MCA,
      3rd Edn., 1972 (withdrawn 1980).

16.   CHETAH (Chemical Thermodynamic And Hazardous energy release
      evaluation program), Philadelphia, ASTM, 1974.

17.   REITP2 program, T. Yoshida *et al.*, <u>Chem. Abstr</u>, 1980,
      <u>93</u>. 188632.

Discussion

Q1   Is the information system that you showed a
                    labelling system or is it designed for use as a data
                    sheet? If it is going to be used for labelling is
                    it not going to cause a lot of confusion?

Bretherick - It is in fact being introduced as a labelling
                    system.

Q2   Could Mr Bretherick give any further details of
                    the US/Japanese computer programs to predict
                    chemical hazards and interactions?

<u>Bretherick</u> - I only have time to do so very briefly. They
            are based on thermodynamic calculations from the
            chemical structure. The maximum energy that
            could be released in the situation concerned is
            calculated. They then go on to assess whether
            there is a high, medium or low risk of explosion
            <u>etc</u>.

Hazards of Apparatus, Equipment, and Services

By Dr I. Szentpéteri

DIRECTOR, CHEMICAL INDUSTRIES ENGINEERING CENTRE, BUDAPEST, HUNGARY

The rapid development of chemical processes has given rise to an increase in the amount of research and experiment performed in laboratories.

In this development the equipment of laboratories has not always corresponded to the new requirements because sometimes the new safety requirements had just been concluded from the experiments performed in the laboratories. It is also a fact that a widely exercised concept exists. We heard it also here stated that in order to hasten work in laboratories equipment is used which the researchers in laboratories would protest against if applied in production lines. This is because the number of highly qualified people in laboratories is large, and the assisting staff are also properly qualified.

The above opinion ought not to be accepted, but it has to be taken into consideration very frequently. In my experience, when the deficiencies of an apparatus are known, it is possible to prepare for the elimination of the hazards. If the hazard cannot be excluded completely there is always a way to minimise it.

I would like to tell you about such a situation. It happened about thirty years ago in relation to the oxidation of ethylpyridine. The reaction was so vigorous that the reaction mixture boiled over from the flask and usually ignited. Time was short and we wanted to continue the experiment - we had no other choice but to work with a 10 l flask with a man with a powder fire-extinguisher at hand. Of course, the reaction was under strict control and direction when the production process was put into operation.

In another lucky instance, which also occurred many years ago, one of my colleagues was working with a small autoclave (its capacity was 200 ml). Perhaps it is an overstatement to call this equipment an autoclave, as it had only a 6 mm wall thickness, and no safety valve. My colleague asked me to take care of the temperature of the equipment which was warmed in an oil bath, while he went away for a time. I had no problem with the temperature which remained constant. However, much more of a problem, as was realised later on, were the preliminary calculations. According to the calculations the pressure should be 2-3 bar at the amidation reaction, but it actually went up to 30 bar, as the correct calculations showed later. Consequently the equipment exploded, and it was only by luck that nobody was injured.

In another experiment we could not heat a furnace to the appropriate temperature owing to the low thermal value of the town-gas. In order to increase the thermal value of the gas, we passed it through a washing flask filled with benzene where a carburisation should have taken place. The burner, however, was not up to the necessary standard and the gas burned back causing an accident.

During a laboratory experiment a gas containing 80% by volume of oxygen was compressed with a compressor which was provided with oil lubrication. Although the compressor had an oil separator, which separated the oil drops, an oil mist could not be avoided. A dangerous situation therefore existed since the oil mist might explode in the presence of a high oxygen concentration. The experiment was discontinued rather quickly.

It is a frequent problem in relation to the hydrogenating reaction, that the applied explosion-proof motors are not the type that would provide proper protection if hydrogen were also present. Furthermore hydrogen tanks are often not placed in separate rooms.

An uncommon accident happened in an experimental laboratory owing to the application of a certain ball valve.

The ball valve in question had been used earlier in another
experiment. Surprisingly an explosion took place at this
ball valve in spite of the fact that it was only installed
into a steam line. Afterwards the investigation revealed
that in the previous experiment the ball valve had been
installed in a material line, which decomposes when heated,
and probably the dried residual material in the valve had
begun to decompose and caused the explosion.

In earlier years it was a custom to establish central
open-plan laboratories, in which senior persons worked. In
one such small laboratory a process involving ether extraction
was performed. About 10 m from the room a gas flame was
burning in the large room. The ether spread on the floor
and then exploded. Only luck prevented the staff from
suffering serious injuries.

What can we conclude from the above examples? There are
two opinions. One is that in the various laboratories current
research may only be performed if the equipment and procedures
of these laboratories differ basically from earlier practice.
This means that the equipment used should comply with higher
safety standards which might then be required in the production
line in a later phase. The other opinion is that it is not
enough to provide safe equipment and supply systems but that
safety considerations should begin with the design of the
laboratory.

For heating in laboratories gas is generally used in
normal circumstances. In this way open flames are generally
present in the room and the staff working in this room are
liable to become complacent about the hazards. As this
attitude is widespread, it is important again and again to
deal with the requirements and possibilities for safe work
in laboratories and hope that the individual investigation
of various fields will reveal the sources of hazards and
the costs of eliminating them.

The first thing to take into consideration is the location
of the laboratory rooms and the relationship between the rooms.

Industrial firms usually placed their research laboratories in high buildings, especially at the time of industrial boom.

This solution, mainly for the first of such buildings, led to an arrangement where the connections between the rooms were made through doors, as the generally required emergency exit was provided in this way. In the case of high buildings a second exit might be technically feasible, through which persons could escape to the open. I have already mentioned the dangers of this arrangement, namely the increased danger of explosion and fire. However, the danger of spreading infection is also greatly increased in biological and hygiene laboratories. (Theoretically, the connection between the rooms may be established with air locks but this does not comply with the requirements of escaping.)

Proper safe laboratory equipment and utilities may only be established in adequately positioned laboratories. So, what would be the appropriate solution regarding laboratory buildings? There are several ways to achieve the correct result. The best of course, in terms of safety, would be to establish only single-level laboratories. However, this is not feasible most of the time because of the limited amount of space available. The other solution is to establish escape or emergency corridors. This solution is rather expensive, but it is unquestionably advantageous, and it reduces the hazards significantly. The escape corridors might sometimes also be balconies, which is - in my opinion - the best solution. In these cases the shading of the laboratories would be solved as well. In the case of high buildings, however, it is preferable to establish closed corridors, mainly for psychological reasons.

The solution of utilities serving the laboratories also may be regarded as a consequence of the architectural solution. Two methods are generally used for the establishment of water, steam, gas, and sewer systems. In one of the methods there is a main line system located in the corridor, in the basement or on the ground floor of the building, from which the laboratory

rooms are supplied through respective shafts. In the other
case lines are established in the central corridors of every
floor level and the respective laboratory rooms are supplied
from these lines. It is clear that the latter method is
better in terms of the above-mentioned requirement to make
the room independent. Lines in the open are easier to
establish, and the sewers need not be placed under the floor.

As far as the electrical system is concerned it is also
more advantageous to supply the respective rooms independently.
The situation is not so simple in the case of fume cupboards
and sewage waters, where the hazard sources are larger.

A complely independent system of the fume cupboards of
the laboratories is the basic condition for the safe operation
of the laboratory complex. There are known accidents where,
because of the connection of fume cupboards, gas infiltrated
into   another laboratory and caused injury. Such an arrangement
also causes problems with ensuring air and temperature balances
of the rooms containing the cabins.

My own opinion is that it is advisable to establish
independent exhaust systems for the fume cupboards even within
a laboratory. In this way the work in the laboratory could be
made flexible as well. The realisation of these independent
exhaust systems for fume cupboards is limited by the height of
the building. The vertical cross-sections might be significant
in tower buildings. An additional problem is to select the
appropriate material for the ducts because these may have to
operate in a very corrosive environment.

It is evident that the expense of these 'exhaust chimneys'
would be worth while if the operation is safe.

I have intentionally omitted the problem of air-conditioning
of the laboratories as this is not generally used in Europe. In
the case of this unquestionably up-to-date system the air blowing
from the central unit causes no problem at all. However, the
exhaust is rather troublesome as it must be done separately.
Although it is general practice in laboratories to perform

smelly and dangerous operations in fume cupboards, we cannot
prevent solvents and the vapours of toxic agents from getting
into the air of the laboratories. For this reason exhausted
air cannot be recirculated. It has to be deodorised and
dusts and dangerous material must be removed from it with
various filters before venting it to the open air. Owing
to the high energy costs it is advisable to use the heat
of the exhausted air for warming the fresh air to be blown in.

I realise that the architectural and engineering solutions
to the requirements listed above are expensive, but without
these means one cannot imagine an up-to-date laboratory
corresponding to present knowledge.

It is not easy to remove sewage water safely either. The
preferred solution would be to establish independent sewage
systems for the respective rooms, but generally this cannot
be managed owing to the limited amount of available space. In
this way the rooms being one over another are connected to the
same vertical drain. In this case every connection must be
provided with a siphon water trap, which does not provide
complete separation, but does undeniably improve safety. I
would emphasise the importance of establishing water-traps at
every connection, as the dangers resulting from the sewage
water may be reduced in this way.

I have not yet mentioned the community facilities, only
the technological requirements of the engineering systems.
The engineering systems of the community facilities should
be independent of the technological systems.

I have also not mentioned the hazards in laboratories
using radioactive materials. If the radioactive material
used in the laboratory is low level, then the sewage water
should be checked for radioactivity before releasing it into
the utility sewer.

In the field of electrical equipment many opinions exist,
especially in relation to research laboratories. The discrepancy
is obvious: open flames are occasionally used in laboratories,
yet the lighting, switches, and socket connectors have to be

explosion proof. However, this is not really a contradiction.
It is evident that it is necessary to provide for occasional
working in the laboratory using gas flames. At the same
time, safe working conditions must be maintained in the presence
of solvents and the electrical system should be established
accordingly.

Although it is not general practice today to work with
laboratory equipment having instruments that indicate the
presence of solvent vapours above a certain concentration,
we do encourage the trend towards this. In spite of the
different calibrations the instructions can be connected
to the ventilation system which further increases safety.

I would like to propose to take the special safety
requirements of laboratory apparatus and equipment design
into serious consideration. I have used the term. 'I would
like to propose', which I think requires some explanation.

First of all I would remind you that we are concerned
with laboratories and not with some kind of 'semi-production'
situation. (In the production situation we cannot allow any
loose ends.)

In recent years laboratory instruments and equipment have
been introduced, which comply with the highest safety standards.
Whereas we have to count at least 50 years of use and wide
applicability of supply systems in laboratory buildings the
instruments and equipment are used for special projects. As
the equipment having the highest safety capacity is expensive,
it should be used where the continuous research activity
requires it. (For instance, in a laboratory where acetone
or ether is continuously used the equipment should be the
safest type available.)

In cases where it can be assumed that the concentration
of solvent vapour in the air will not rise above 20-40% of
the dangerous concentration, then equipment can sometimes be
used which only partially fulfills the fire and explosion
protection requirements. Of course appropriate and increased
engineering cautions must be applied - including the above-

mentioned instrument for indicating the concentration of
solvent in the air.

I think that the dangers arising from the application
and use of glassware are not appropriate to the present paper.
The hazards of breaking glass are reduced by technical developments mainly by the increased use of ground joints which have
reduced plug manipulations.

Another paper with the same title as this one might
well have included the dangers resulting from the deficiencies
of the clothing of laboratory staff. In this area the level of
correct behaviour is below that required, and essentially
insignificant accidents can become more serious than need be.
For example, in one case a lethal accident resulted from a
minor laboratory fire because the victim's underwear was of
artificial fibre and adhered to the body thus sealing a major
part of the victim's body from the air. This would have not
happened if cotton underwear had been worn. This problem
is not altogether solved and the socks and underwear made of
artificial fibre continue to present serious hazards. The
otherwise very sophisticated BDR standards involve only
recommendations in this respect.

It is impossible to speak about all aspects of this
subject in the short time available. For this reason I have
tried to draw your attention to those aspects which are most
important for studying the various problems of laboratories.
We must recognise that safety in laboratories begins with the
building, then the services and equipment.

Finally, in the research laboratory of one factory that
I visited recently I saw plastic bottles quite close to the
electric heater. Furthermore not very far from the electric
switch there were disorderly cables, wires, and pipes, as well
as a small autoclave. Remember that this was in a research
laboratory. Nor was it an isolated case; I have recently
visited two other laboratories and seen very similar situations.

In the circumstances we must continue to talk about safety until we realise our aims.

No Discussion.

Managing People

By Mr E. Thompson

MANAGER, SAFETY, FIRE AND SERCURITY BRANCH, BP RESEARCH CENTRE, SUNBURY, UK

The two previous papers dealt with the materials and equipment used in laboratories and this paper deals with that most complex piece of equipment, sometimes capable of totally irrational action - the staff. The main purpose of this paper is to describe how the control and supervision of safety in a research environment is carried out at the BP Research Centre. This in no way suggests that these procedures are the only ones or the best but, simply from our excellent safety record, that they appear to work.

The paper first describes the research environment at the Research Centre and then discusses the basic differences in approach to safety matters between research and production sites. The safety training of a new graduate is described and, finally, the range of safety advice available and the controls that apply to a fully professional chemist. The overall policy is to develop a system which encourages professional behaviour, self analysis, and a willing co-operation with safety procedures.

The Research Centre

The BP Research Centre occupies a site of some 35 acres with a high density of building. There are approximately 1600 staff, of which 650 are graduates. The Centre is the principal research establishment of the British Petroleum Company plc and covers a very wide range of experimental topics, many outside the traditional interests of the oil industry.

The Research Centre is organised on a Divisional basis. This is a conventional arrangement for any large organisation. The varied activities carried out in the main research-based

Divisions involve a range of hazards that is equally wide.

The graduate staff cover a considerable spread of
scientific and technical disciplines. Some are present
as specialist advisers or groups, although the majority
fall into the broader chemistry, physics, and engineering
categories.

The Research Environment

The next point concerns the basic differences in safety
philosophy between factory and research. In the oil industry
this is often the difference between refinery and research
station. In a refinery the hazard is obvious and stems from
the enormous hydrocarbon burden in the tanks and in the
processing system. Hence, for a refinery or factory the size
of the hazard is so substantial as to require strict adherence
to a code of behaviour with severe penalties for transgression.

At the Research Centre the total tankage is a mere 190,000 l,
the hydrogen supply bank 0.25 tonne and the liquified petroleum
gas(LPG)supplies 15 tonnes. Therefore, the actual hazard from
explosive materials is much less. However, in contrast, the
Research Centre has a very much wider range of activities, e.g.
radiochemistry, genetic mutation, partial oxidation, and high-
pressure reactions, all in a concentrated area. The fundamental
difference is that the whole object of the work is to investigate
untested, novel problems of considerable complexity and unknown
response. Under these conditions the imposition of too rigid
a safety code would be inappropriate.

Under factory conditions the management probably has, out
of necessity, a very hierarchical system with limited
independence for individuals apart from the senior management.
The research situation is exactly the opposite. In the research
environment the hazards are lower but the wide range of activities
and the need to encourage independent action place considerable
dependence on the ability of the individual to act in a safe
manner without supervision. Thus one is not likely to have a
Flixborough disaster but a series of smaller, laboratory
accidents which, nevertheless, can be serious and often quite
difficult to resolve. A flexible approach is needed and indeed
it will be the experience of anyone who manages a large number

of graduates that any direct attempt to introduce a dogmatic
type of rule will often be met head-on with resistance. The
combined intellectual opposition of large sectors of the
graduate staff can be formidable. Persuading them on rational
grounds to your way of thinking is significantly better but
does require an in-depth understanding of their problems. In
turn they must have respect for the safety staff. Thus it is
essential to have safety personnel capable of being accepted
as technical equals by the chemists, rather than being seen
as thoughtless imposers of irksome regulations.

The next part of this paper is devoted to a description
of the safety rules and systems as applied to new graduates
and traces the progress of a new graduate through the stages
of recruitment and early research training.

The New Graduate: Training

For the graduate, the first point of contact is the recruitment
interview. It is the author's opinion that there should be some
safety input even as early as the selection procedure. However,
it requires an interviewer of exceptional perspicacity to detect
the inept, clumsy, and dangerous graduate by interview. It
would, perhaps, be of benefit if stricter practical and safety
training were carried out as part of the university curriculum.
In fact the university safety picture appears to have improved
considerably over, say, the past 10 years. An important example
is that few graduates now object to wearing safety glasses in
laboratories whereas a few years ago such an elementary
precaution was unheard of. Universities and other institutes
should be further encouraged by industry to treat safety as
an essential part of a fully professional attitude.

Having recruited the graduate the safety policy is simply
based on the philosophy that 'ignorance of the law is no excuse'.
While this puts the onus on the graduate to read the rules, it
also puts a considerable responsibility on the company to make
sure that the 'law' is clearly explained and that training is
available. As a first step, the graduate is given lectures on
the following basic subjects:

(i)   Fire - building evacuation and laboratory equipment;
(ii)   Safety and health;
(iii)   Safety manual;
(iv)   Project reviews;
(v)   Fire fighting and breathing apparatus;
(vi)   Specialist courses.

These lectures and demonstrations are fairly standard in
any organisation and it is not proposed to describe them
further. The Safety Manual is, however, a key document and
worth further consideration. It is the basic rule book and
the book to which common reference is made in all disputes
regarding safety matters. It contains about 180 pages in 23
sections. These sections cover almost all aspects of laboratory
and workshop practice, (see Tables 1 and 2).

Table 1   Safety manual
Policy: Safety and Health Organisation

Precautions	Equipment
Permits to work	Pilot plant
Injuries and reporting	Laboratory apparatus
Fire and evacuation	Workshop machines
Accident prevention	Engines - test rigs
Protective clothing and equipment	Lifting equipment

Table 2   Safety manual

Hazards	Techniques
Laboratory	Low-temperature work
Dangerous materials	Waste disposal
Glass	Microbiology
Radioactivity	Gas cylinders
Electricity	Isolation of vessels
Biochemical	Welding
	Contract labour
	Building services
Signatures	Site maintenance

These rules have been developed, considered, polished, and changed with experience over many years. It has proved a very successful document and is in demand throughout the BP Group. In the author's opinion, its success as a reference document follows from the fact that, because the rules are 'live' and are developing with time and experience, they are accepted as correct current practices by the technical staff and are not seen simply as immutable tablets of stone. As can be seen from Tables 1 and 2 the sections cover such diverse topics as microbiology procedures, low-temperature laboratories, and design of pressure vessels. Each graduate is instructed to read the sections appropriate to his work and sign a register. In addition the Research Divisions often have simple one or two page notes on local safety matters of importance. Thus the new graduate is lectured and trained in the essentials of safety and given a set of rules. As a reminder, all technical staff are required to re-read the rules every three years.

This is clearly only part of the story, for no safety control can exist without adequate job control and management supervision. A basic division structure is shown in Table 3.

Table 3   Divisional organisation

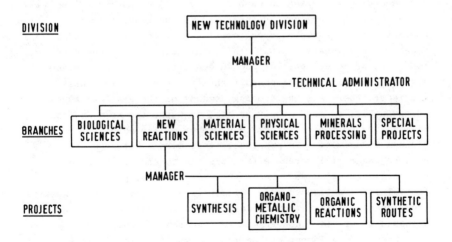

The new graduate is organised into a Project and his immediate
superior is a Project Leader, who is considered to be an expert
in his particular field of activity. As far as possible the
Project Leader will allocate a new graduate a specific job in
the project programme. This definition of a job is important
at this early stage for it helps the graduate to limit his
boundaries of operation more easily and, equally important, it
enables the Project Leader and senior chemists to monitor his
progress without obvious interference. Clearly not all research
topics can be fragmented to provide exactly one-unit operations
but, wherever possible, this is attempted.

The New Graduate: Safety Support

Thus, in this early phase of his career the young graduate
has several people to call upon for advice: his immediate
Manager, his Project Leader, his colleagues, assistants and
operating staff, safety staff and representatives. The Project
Leader, safety staff, colleagues and assistants are all obvious
contacts but there is, in addition, an Area Engineer, an
important contact which deserves further explanation.

The Research Centre is divided into areas, partly
geographical and partly activity controlled and in charge of
the general engineering in each area is an Area Engineer. He is
an experienced practical engineer and is the graduate's main
contact with workshops and engineering services of all kinds.
Clearly the graduate can build his own equipment but where he
requires, or more important should require, expert engineering
advice he is expected to consult the Area Engineer for these
services. Where the engineering demands of the graduate appear
unreasonable, or more likely unsafe, or both, the Area Engineer
will refer the matter either to a member of the safety staff or
his own management. In practice, it is more likely that the
Area Engineer will persuade the graduate to see the error of
his 'engineering' ways. Hence there is a local, positive but
tactful and experienced safety limitation put onto the engineering
requirements of the graduate chemist. A further restriction,
although in practice no barrier to research, is that no
high-pressure equipment can be made or modified in his Area
without the Engineer's approval. We have found in practice that,

with an experienced engineer, an easy rapport soon develops
between the engineer and other disciplines. With growing
confidence the graduate will accept his advice and design
assistance without argument and actually look to him to protect
his interests in workshop queues and in obtaining design
facilities. The Area Engineers, on the other hand, seem to
have no difficulty in the dual role of providing engineering
advice and at the same time maintaining engineering standards.
Should a serious dispute develop he, like the village constable,
tends to stand to one side and allow the issue to be settled
either by the safety staff or the engineering management.

The Safety Representative was also mentioned as assisting
the graduate. This is a post required as part of the Safety
Committee procedures. At the Research Centre this is usually
an experienced senior chemist who represents the branch or the
laboratory on a Safety Committee and is the local focal point
for safety enquiries.

Finally the role of assistant or operator is not to be
ignored. Although they are not in charge, it is, in the author's
experience, a poor assistant who will not give direct,
frequently unsolicited, and almost always good advice on the
operation of equipment. Likewise, the graduate's colleagues are
frequently first to point out if there is something wrong with
his experiments, often with considerable candour. The final
check in the early stages of the graduate's career is that of
the annual staff report prepared by his Project Leader and upon
which his promotion depends. This often does, and in future
certainly will, make a comment on his experimental ability with
particular reference to his safety sense and the protection of
his colleagues.

From Table 4 it can be seen that with his Project Leader,
colleagues, assistants, Area Engineer, Safety Representative,
and eventually safety staff observing and controlling him, the
graduate's life may seem to be totally prescribed. In fact all
these controls operate with a light hand. They seek to boost
the graduate's confidence to attempt new things with the
assurance of help when needed and rescue if something should go
wrong.

Table 4   <u>Safety support for the new graduate</u>

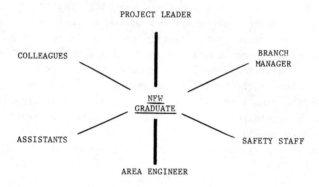

This 'graduate' instruction phase can be summarised in
the following stages:

- Take as much care as possible in recruitement;
- Know the law - ignorance of the rules being no excuse;
- Have a sensible rule book;
- Basic training by courses;
- Sensibly limit the range of job;
- Provide safety assistance - which must be used;
- Positively record his progress;
- Attempt to inspire confidence in safety staff.

The Experienced Chemist

Following this first phase, which can take a number of years,
the graduate's ability improves through experience to a point
where he can operate either entirely alone or in his discipline
in a multi-disciplinary team. An average of 4-6 years, sometimes
longer, may be required before he is considered capable of
organising and controlling his own work and that of others.
The remainder of this paper deals with the safety cover as
applied to a fully professional chemist.

Management and Administration

From a safety viewpoint the management and administration
differences between the graduate and the early years of a

professional chemist can be summarised as follows:

1.   The chemist's programme will be wider either in
      real activity or in technical and commercial
      importance and he can become a junior project leader
      himself in a restricted area.

2.   His authority to purchase chemicals, equipment, and
      services will considerably increase with only limited
      checks.

3.   He will be encouraged to be an active member of the
      division or branch Safety Committee, either as a
      general member or as a specialist in a particular
      field.

Point (3) leads to the Safety Committee structure required by
legislation and defined by company policy. There are two
levels of committee, the Research Centre Committee and the
Division Committees. The Division Committee membership is
arranged so that all laboratories, workshops, offices, and
kitchens have representation. These are the local Safety
Representatives with other elected members of staff. The meeting
is chaired by the Division Manager and the topics discussed
are principally within the specialisation of the Division's
work. The senior committee is the Research Centre Committee.
It consists of the General Manager and Divisional Managers
with elected Division representatives. This body discusses
important Research Centre items, any serious accidents and then
investigates and sets policy with BP Group guidelines.

The key points of the arrangement are first that every
member of staff has ready access to a representative and, if
necessary, can himself present a safety point, and secondly
that the senior management of the Research Centre can be seen
to endorse safety practices and, in fact, to play a very active
part.

Plant and Equipment

Because of the high severity of operations in terms of
temperature and pressure of much of the equipment, we have
to establish strict control over engineering standards.
From a chemist's point of view the procedure is as follows.

He first discusses the matter with the Area Engineer
and possibly a design engineer and produces a polished
version of his own 'back of an envelope' sketch. A design
dossier is put together and presented to the safety staff.
The design is discussed with the chemist and engineer and
checked for any toxicological or operational problems that
may be encountered. The scheme is then costed, capital applied
for, and formally sanctioned. The detailed design is then
completed in conjunction with the chemist and finally approved
by the engineering services management as conforming to the
correct engineering standard. Any hazard analyses are carried
out at this point either by other senior members of the project
or by the engineering and safety staff. Plant and equipment is
then constructed under the supervision of the Area Engineer or his
deputy and checked by the chemist. If the plant is of pilot
size an operating certificate is required. This is a document
issued by the safety staff following a detailed examination
of the plant and, more importantly, the proposed operating
procedures. Hence there are two design and one chemical
checkpoint, one hazard check, and one operational check. The
system also provides a record of the design philosophy and
decision.

However, it is in the nature of research that having
installed this vital piece of equipment it will probably not
work. We do not prevent the engineer or chemist from
modifying equipment, and obviously changes will often be
required. With bench scale this is no problem; with larger
sizes or more complex equipment the Area Engineer or a member
of his staff carries out modifications. Where equipment is
designed to operate above 140 bar it is registered as a
'designated' unit. On these units only the chartered engineers
and craftsmen are allowed to make any alterations.

Occupational Health

Sunbury is host to the Group Occupational Health Centre,
which provides a service to assist our chemists in
occupational health, medical, and toxicity matters. This
organisation, controlled by senior medical staff, advises all
staff, not only at Sunbury but throughout the BP Group on
occupational health matters. Any problems regarding the
toxicity and short- or long-term effects of materials can be
taken up with this group. They also maintain up-to-date
medical advice and advise on new standards and controls and
issue guidelines on procedures in potentially dangerous areas.

Work Abroad

Special problems arise when the research work is off-site.
For instance we have groups in the Arctic and several in the
Middle East. A chemist will need and receive specialist local
advice from operating departments, safety staff, and his
management. Special care is needed in the selection of staff
and safety equipment where operations have direct contact
with the general public.

The safety staff have a responsibility to ensure adequate
safety cover for staff on location. In the UK and Europe
this poses no serious problem but can be difficult in Alaska
and parts of the Middle East. This is an area where
inter-company and national co-operation is often needed to
provide adequate services in an isolated location.

Reporting Systems

We expect every accicent on the site, however minor, to be
reported to full-time medical staff who make out a treatment
report. The majority of these reports cover trivial injuries
which account for something like 99% of all accidents and are
of a similar nature (such as cuts, burns, and bruises). Copies
of the treatment report are sent to the Safety Office and any
'suspicious' cases are investigated by safety staff. In the
case of more serious injuries or more important near misses, an
accident report must be completed, and this is circulated to
the management and the safety staff. Committees of investigation
are automatically set up where the accident or 'near miss' is
serious.

Safety Audit

One of the most important self-controls that the chemist will encounter is the Project Audit. This is a local audit of a project area of operations by its own members. It is a Research Centre rule that these critical assessments of their own operations have to be carried out on an annual basis for large projects and more frequently for smaller jobs. Often the specialist members of other projects may be called in for advice. The whole of the operation of the project is critically examined by its members and, most important, their findings are reported to their own management and the safety staff. This has been found to be a most effective self-regulating system especially as each member of the project has to defend his own operations to his colleagues.

During this final development phase in which the chemist will reach full professional activity, the advice available to him is as follows:

1. Chemistry - Company specialists, consultants, and safety staff.

2. Medical, Toxicological and Environmental - Group Occupational Health Centre, Medical Officer, and Environmental Control Centre.

3. Mechanical - Engineering branches, Area Engineers, design engineers and safety engineer.

4. Legal - Safety staff, Administration, Legal Department and Group Safety Centre.

5. Legislation - Safety staff and Group Safety Centre.

The role of the Safety Officer (whatever his local title) has been mentioned at various points and his activities are obvious in most cases. The author's opinion is that to function effectively all the safety staff must have sufficient authority to carry out any decision without undue argument, although such authority must be used as a last resort. The Safety Officer

must establish sufficient credibility and professional
reputation to encourage staff to consult him willingly.
This essential rapport can be lost by any tendency towards
autocratic or draconian application of the rules.

In summary there are three simple philosophies:

- Attempt a consensus view of safety rules;

- Put the onus for safety control on the chemist's
  professional ability;

- Provide proper professional safety services
  and encourage their use.

Finally, and most important, all scientists must be
encouraged to communicate with their neighbours for, in many
cases, a man's best safety officer is his colleague.

Discussion

Q1   Mr Thompson showed a slide of a pilot plant at
          Sunbury. Could he tell us whether the plant
          is flameproofed?

Thompson - No, that particular plant is not. We are moving
           towards zone 2 areas for all pilot plants and
           new laboratories are being made up to zone 2 -
           certainly not zone 1.

Q2   Who does Mr Thompson report to? Does he consider
          this a satisfactory reporting relationship and
          who appraises his performance?

Thompson - I report to the General Manager for general
           operating purposes. We have a group safety
           organisation which covers the whole BP group and
           my performance is watched over by this and by the

General Manager. My annual report goes to the
group safety organisation which incorporates
it into the overall statistics.

Q3   Mr Thompson has demonstrated the advantages of
            large organisations in the field - especially
            for training graduates. Do you see any advantages
            for smaller companies in outside training schemes?

Thompson - There are one or two areas where an external agency
            might be useful; certainly in the case of new
            graduates.

            Occupational health is becoming increasingly
            important to us - smaller companies must find it
            very difficult to keep up with monitoring
            requirements etc. without an occupational health
            officer.

            The fire service and the HSE are very helpful in
            some areas but in my opinion it would still be
            worth duplicating their advisory services.

Q4   Can you tell me how your performance as a safety
            manager is evaluated. For example, what specific
            criteria, if any, are used?

            This is causing great difficulty in the US.

Thompson - This is obviously very difficult (except in
            extreme cases). It is especially difficult on
            the research side because of the support that is
            required by the staff - the hazards and the
            requirements are obviously different to those on
            a production site.

            Perhaps the most important quality in a safety
            manager is the ability to persuade people to do
            things properly. The ability to anticipate
            problems and to administrate is also very important.
            I don't think I can really add to that.

What Standards Should We Use?

By Mr T. Rose
THE OFFICE OF FAIR TRADING, LONDON (FORMERLY OF THE HEALTH AND SAFETY
EXECUTIVE, LONDON*), UK

This paper considers the underlying legal framework and 'standards'
in use in the United Kingdom (UK) for the control of toxic
materials, and from that background develops the direction for
the future. The paper also makes reference to monitoring and
its relevance in the overall control of risks.

Risk Assessment
Risk assessment brings together two elements - the consequence
(or outcome) of an event and the probability of that event
occurring. In occupational hygiene matters the outcome can
often be related to the exposure level. The exposure level at
which a gas (e.g. carbon monoxide) kills is generally well known,
and exposures at or in excess of this level will kill: the outcome
is death with a probability of 100%. Similar outcome/exposure
relationships can be developed for unconsciousness, narcosis,
irritancy, etc., with a reasonable level of certainty that the
exposures will lead to the expected outcome. The exposure levels
are often based on medical and scientific observation at the time
of incident or in the immediate follow-up investigation. This is
possible because these acute effects can be directly related to
exposures in the immediate past.

I believe that the main concern and difficulty lie with chronic
diseases, expecially those with long induction periods. Chronic
diseases are much less easily related to exposure levels. There is
no equivalent exposure level for cancer, as there is for gassing,
at which, if it is exceeded, there is high level of certainty of
effect. More importantly there may be no exposure level at which
we are certain that there will be no effect. A safe level cannot
be prescribed.

*The views expressed in this paper are those of Mr Rose
and are not necessarily those of the Health and Safety
Executive.

Where there are or have been problems, good documentation
of the individual exposure levels over long periods before the
onset of the disease is generally not available. Estimates
therefore have to be made from other sources, usually by
extrapolation from some specific study or from animal data.
The basis of any attempt to relate outcome to exposure levels
is only as good as the data and often the data are
flimsy or unavailable. Nevertheless some attempt to define
levels which might be harmful has to be made and this is the
approach of the Threshold Limit Value list produced and
documented by the American Conference of Governmental Industrial
Hygienists. Until recently this list was published in the UK
by the Health and Safety Executive HSE in its Guidance Note
Series.

Threshold Limit Values

I would suggest that 8 hour time-weighted average Threshold
Limit (TLVs) have little relevance to laboratories - where
workers are unlikely to be exposed for continuing periods.
There are perhaps some exceptions in factory quality control
laboratories, or perhaps where specific solvents or mercury
are constantly used. I believe that any laboratory which is
exceeding the 8 hour time-weighted average would have to be
very sloppy indeed. Exposures in laboratories approaching the
time-weighted average TLV would perhaps be an indicator of
bad laboratory practice rather than reassurance that everything
is satisfactory. Short-term exposure limits and Ceiling Values
are associated with more acute problems and may be relevant to
laboratory exposures but are not relevant to chronic effects.
TLVs may not be available for many of the materials found in
chemical laboratories. Therefore the use of TLVs is limited
and not particularly satisfactory for chemical laboratories.

The Law

Work activities in chemical laboratories are subject to the
Health and Safety at Work etc. Act 1974 (HSWA). Section 2(1)
of the HSWA places a general duty on employers to their employees,
and Section 2(2) contains the matters to which the duty extends.
Throughout the Act runs the phrase 'so far as is reasonably
practicable'. Understanding of these words is important in

practical application of the law.

A very important case in the interpretation of 'reasonably
practicable' is Edwards v. National Coal Board 1KB 704(1949),
where Lord Justice Asquith said: "Reasonably practicable" is
narrower than "physically possible".......... and implies that
a computation must be made in which the quantum of risk is placed
in one scale and the sacrifice involved in the measures necessary
for averting the risk (whether in money, time, or trouble) is
placed in the other and that if it be shown that there is a gross
disproportion between them the risk being insignificant to the
sacrifice - the defendants discharge the onus upon them'.

This interpretation very clearly calls for the balancing of risk
against control measures and puts the duty with the employer.
This duty is re-emphasised in Section 40 of the HSWA.

Lord Justice Asquith continued 'Moreover the computation
falls to be made by the owner at a point in time anterior to the
accident'. The principle being established is that the 'balance'
can only be made on the basis of facts that were known or that
ought to have been known to a reasonably well informed employer
at the time and the risk is foreseeable from the nature of the
operation.

The phrase 'as far as is reasonably practicable' therefore
requires consideration of control measures. At this point I
would like to refer back to the basis of TLVs, i.e. biological
effects only with no 'control' considerations.

Control Limits

In the UK , Control Limits, which bring together the biological
and control elements, have been introduced for some substances
(e.g. asbestos, lead, and vinyl chloride).   The procedure
for development and adoption of Control Limits in the UK is worth
noting. Control Limits are recommended to the Health and Safety
Commission by the Advisory Committee on Toxic Substances, which
consists of representatives of employers, employees, local
authorities, and scientific and medical experts.

The Committee develops, from all the information available,

a level of exposure where the risk is acceptable, and further
expenditure of effort in those processes where control is most
difficult is out of all proportion to the reduction of risk.
The legal obligation, however, is to reduce still further in
those processes where it is reasonably practicable to do so.
The Control Limits represent an upper limit of permitted
exposure; they do not represent a safe level at which no
further control improvements are necessary.

I believe that this approach, which links together control
and biological evidence, brings limits which are widely accepted,
sensible, and have .credibility. There is nothing worse, from
an enforcement point of view, than standards which have lost
credibility with all those who work with them.

The development of control limits is relatively slow,
it will certainly never cover all materials, and in most cases
the limits will be based on day to day, 8 hour exposure. It is
likely therefore that these standards will still not fit most
laboratory situations. I would suggest that no general exposure
limits will suit laboratories with a wide range of chemicals,
many of which may be at an experimental or development stage.

Where Do We Go for Standards?

If the 'standards' set by the 'external agencies' are not
relevant or applicable to chemical laboratories - who does
set them? The case law suggests that the responsibility lies
with the employer. S.40 of the HSWA also specifically points
out the responsibility of the employer.

The person in charge of the experiment/laboratory should
have detailed knowledge about the materials being handled and
the hazards involved. There is something seriously wrong with
the situation if they know less than the Inspector from the
enforcement agency. The person in charge should not only have
the detailed chemical expertise himself but often has colleagues
or access to others within the same organisation who have
complementary skills (e.g. toxicological, medical, and biological).
It is from this base of expert knowledge that standards for the
control of the whole laboratory or specific protocols or conditions

for higher-risk experiments are developed. What sort of
knowledge is needed to set standards? I would suggest:

Knowledge of materials	– Chemical properties (e.g. reactivity, flammability)
	– Physical properties (e.g. gas, liquid, dust)
	– If the material is new, what do we know of analogues?
Knowledge of equipment	– How are materials handled? What problems are likely?
	– Limitations of containment (e.g. centrifuges, fume cupboards, etc.)

Knowledge of people and their limitations.

From this background knowledge, and detailed knowledge of
the experiments and the materials, protocols can be developed
for the safe handling and use of the chemical. Different
protocols are obviously needed when using materials which
are for example, toxic by ingestion, absorbed through the skin,
or possibly allergenic.

The 'Expert' Safety Committee

Biological safety committees have been used by those organisations
involved in genetic manipulation work to assess risks and develop
work standards and protocols. These committees have a different
role from normal safety committees, which may consider more
general matters. The expert committee should consist of a
combination of those who have detailed knowledge and representatives of those who will carry out the work. It can be a forum
not only for standard setting but also for the exchange of
information. I see no reason why chemical laboratories should
not have similar types of organisations for setting standards
as new experimental programmes are developed, or new chemicals
are introduced into the laboratory. The normal laboratory
safety committee may be able to handle this work, but it is
likely that special working groups within the overall management/

safety framework may be more appropriate.

The Enforcement Agency (HSE) Role

Having put the onus for risk assessment and the development
of control measures and standards upon the employer, what is
the role of the enforcement agency? I would suggest that it
includes:

- Ensuring that the in-house procedure is carried out
  consciously, conscientiously, and correctly;

- Making certain that the actual arrangements adopted
  are satisfactory and consistent with the in-house
  procedure;

- Providing guidance/information on good practice and
  the prevention of incidents.

The Chemicals and the Education National Industry Groups (based
in Liverpool and Barking, respectively) have responsibilities
respectively for the Chemical Industry and Educational Institutions,
which would cover most institutions for inspection across the
country. The Health and Safety Commission is also advised by
the Education Service Advisory Committee, which considers the
hazards arising from within the field of education.

     In summary, the HSE inspectors' role is to see that the
work is done in the spirit of self-regulation underlying the
HSWA. I personally do not believe that there can be detailed
and comprehensive regulations specific to chemical laboratories.

Monitoring of Control Measures

Monitoring of the adopted control measures is vital to make certain
that they are and remain effective. The monitoring programme must
have the objective of providing relevant and useful information
and the basis for decision making about the effectiveness of
control. The monitoring programme must match the assessed risk
and specifically match the materials and protocols/procedures
involved. For example, there may be no value in air sampling
if the major problem is likely to be skin absorption.

Measurement is a very useful, factual, and objective basis for the monitoring programme, but obviously it is valueless if the wrong parameters are measured or the information is produced at the wrong time. I use the term measurement to include not only the results from air and biological samples, but all other measurements that may be taken (e.g. velocity at the face of a fume cupboard, static pressure measurements in ducts, and filtration efficiency of discharge filters) as part of the monitoring programme. Relevant measurements from many sources can all form part of the information that leads to correct decisions about the effectiveness and adequacy of the control measures.

Measurements to assess the adequacy of control may take many forms. These include:

Assessing personal exposure	– Usually done by air sampling from the breathing zone of the exposed worker for an appropriate time period. Biological measurements may also be used; these are of greatest value where effects can be identified before there is damage (e.g. blood lead measurements)
Measuring equipment performance	– May be taken on a continuous (e.g. fume cupboard instrumentation) or periodic test basis (e.g. space centrifuge containment with test gas or spores)
Work procedures	– Can be done by other relevant measurements (e.g. surface swabbing or settle plates).

The measurements are an integral part of the control of the risk

and the right package of monitoring/measurement procedures
should be developed to fit the risk. Where external standards
are not available, in-house standards (or norms) can be developed
for interpreting the results of measurements and identifying
changes in the control performance.

Conclusion

In conclusion, what standards can we use?

1. Laboratories should be setting their own standards
    from a position of knowledge of the risk. This is
    in the spirit of self-regulation, and a requirement
    of the HSWA.

2. The organisation should have clear procedures for
    setting standards; this is perhaps best done by
    involving those with most knowledge e.g. through
    'expert' committees

3. Threshold Limit Values and Control Limits have limited
    application to laboratory work, but can provide
    guidance.

4. Monitoring programmes which are relevant to, and
    balanced against, the assessed risk are important
    for determining the effectiveness of the controls.
    Measurements can be an important part of the monitoring
    programme.

5. The enforcement agency (HSE) can provide guidance and
    assistance from wide experience as 'standards' are
    developed to suit particular risks. The HSE much
    prefers to be involved in discussing and avoiding
    problems than in clearing up after them.

No Discussion.

Conflict of Safety Interests with Legislation

By Mr N. H. Pearce
UNIVERSITY SAFETY OFFICER, UNIVERSITY OF BRISTOL, UK

In the study of a chemical process it is a standard practice
amongst kineticists to attempt to study the individual
reactions that occur in isolation, the assumption being that
no interactions take place. Although this is a convenient
method of examination it is fully understood that in the real
process these reactions are not isolated but have profound
interactions. In order to understand the whole process, such
alternative, competing, or consecutive reactions must ultimately
be examined and accounted for in the general understanding of
the process.

Modern technical legislation appears to have proceeded,
in many cases, half-way down this path. Unfortunately, it seems
to have stopped at the individual reaction stage and the
legislation etc. appears then to be promulgated on the
assumption that there are no interactions with other factors.
Thus, the design of fire protection in buildings takes no account
of the fact that protective devices may affect other environmental
conditions and, for example, actually enhance the possibility of
increasing air contamination. The advent of the Health and
Safety at Work etc. Act 1974 (HSWA), gave hope that in future it
would be possible to take account of interactions and actually
deal with the real world. This was enshrined in the original
Act in such terms as 'reasonably practicable', which allowed one
to examine all factors including cost-benefit and achieve an
optimum control system which took into account all important
parameters. As an early example I would quote the requirements
concerning the storage of poisons, made under the Poisons Rules,
which required storage of such poisons in a locked container.
For most materials this was perfectly sound. However, the poisons
list did include such materials as the alkali cyanides, which are

well known to react with carbon dioxide and moisture to
liberate hydrogen cyanide at a slow rate, and we are all
familiar either directly or at second-hand with stories of
chemists who opened containers of cyanide that had been stored
for long periods of time. Obviously to store such materials
under closed conditions is the very worst procedure that one
should use.

It does seem that the original intention of the HSWA,
which was to legislate through flexible Codes of Practice, has
to a great extent and without any public statement, been given
up for legislation by regulations. More and more of these
regulations appear to isolate their individual remits from the
real world and operate under the assumption that they may deal
with their subject without affecting other matters. It is
difficult to find where the modus vivendi of this trend comes
from. It is certainly not from the scientists or technologists.
It does not appear to come from the lawyers, and most conversations
with the people in the enforcement agencies, particularly at
the 'hard end', suggest that it does not come from them either.
Perhaps the generalist, the civil servant, or the Eurocrat is now
the source of such legislation. I would like to recall a
discussion that took place in a Royal Society of Chemistry
sub-committee concerning a European proposal entitled 'Proposal
for a Directive concerning the assessment of the environmental
effects of certain public and private projects'. After five
in-depth studies of this long and involved document it became
clear that its true aim was nothing to do with the title. It
was, apparently, a device by which Holland hoped to make Germany
clean up the Rhine before it crossed the international boundary.
Although this may have been a laudable aim it really had very
little concern for the possible enactment of legislation against
firms working in Great Britain, which is devoid of such
international boundaries and is washed by the sea. In order to
examine this problem further I would like to consider a number
of cases in which the legislation does not seem to comprehend
scientific reality.

Of the three fibres commonly used in fabrication in this
country, namely crocidolite (blue asbestos), amosite (brown

asbestos), and chrysotile (white asbestos), there appear to
be quantitative differences in their potency as carcinogens.
There is a great deal of evidence to suggest that blue asbestos
is far more potent than the other two and less well defined
evidence that amosite is somewhat more potent than chrysotile.
Rapidly accumulating evidence is tending to blur these
differences. The differences in the activities and potencies
of these materials are shown by the differences in their
permissible exposure levels. These have decreased over the
past few years until some months ago when the accepted levels
more or less throughout the world were $2f\ ml^{-1}$ for chrysotile
and amosite and $0.2f\ ml^{-1}$ for crocidolite. These are
obviously real differences. However, by an enactment of
December 1982 levels in the UK have been modified so that
chrysotile is now $1f\ ml^{-1}$, amosite $0.5f\ ml^{-1}$ and crocidolite
remains at $0.2f\ m^{-1}$. The American levels remain unchanged
at the previous values. Although not wishing to comment on
the relationship between environmental levels and their
clinical effect I would like to comment on the chemical
features of the analysis on which these levels are based.

A known volume of air in which the fibres are contained
is drawn up by means of a pump operating for a minimum period
of four hours across a standard membrane of known area. The
air flow-rate is constant and it is normal to collect the
fibres by presenting the open face of the collector downwards.
The membrane is then removed from the site and transported to
wherever the microscopic examination takes place. The membrane
is treated with solvent to render it transparent and using a
standard grid the number of fibres presented in a hundred
standard areas is counted using a convention in which the
length, the length/width aspects, and orientation of the fibres,
etc. are taken into account. Obviously, this can by no means be
considered a very accurate form of analysis. Assuming a trained,
competent microscopist in constant practice, his own errors in
counting the same sample, particularly at low concentrations,
can be as much as 50-100%. The errors between microscopists
are similarly large. Work being carried out at the Institute
of Occupational Medicine in Edinburgh is concerned with this
particular problem and many of us are participating in the
exercise. We await the outcome of this work with interest.

The clearing process, the actual transport of materials from site, the errors in air throughout, and the actual validity of downward sampling together add up to likely errors so great that whatever the legislation says one must question whether there is any real, practical difference between the three values suggested as permissible environmental levels.   If there is not, it would not seem to be sound practice to establish such levels by legislation.

A considerable political rumpus has occurred in the last two to three years over exposure to formaldehyde vapour. This is another area in which legislation appears to have parted company with common sense. If formaldehyde at low level was as toxic as many of the claimants propound it would seem difficult for any medical doctor to survive his period of training.   We must actually look at the proposals in terms of concentration, etc.:

1.   There appear to be no attempts to differentiate formaldehyde from formalin, although, as most chemists are aware, aqueous solutions of formaldehyde are chemically very different from formaldehyde gas.

2.   The suggestion that the permissible level of formaldehyde should be reduced from 2 ppm to 1 ppm must be examined scientifically. Whatever long series of biochemical reactions that result in the final pathological conditions must occur reasonably fast. Formaldehyde would otherwise build up to an intolerable level in the lung and blood. It seems likely that the rate-determining process is governed by transfer of formaldehyde from air into solution. Since we are dealing with initial fast liquid reactions this must be a first-order process proportional to the partial pressure of formaldehyde. The actual reaction rate is therefore proportional to the concentration differences, i.e. 2 ppm and 1 ppm, even assuming that formaldehyde behaved in an ideal fashion. Since it does not, it would seem probable that there

is very little difference between these two
levels in the activity sense. Certainly when
one takes into account problems of formaldehyde
analysis by simple means at these levels it is
probably impossible to differentiate them.

Some years ago we examined a problem which nearly all of
us have met, i.e. the incorporation of numerous smoke/fire
doors into buildings in which chemical research is being
undertaken. Examining the proposal that each smoke door is
itself a small but finite hazard we obtained a graph (Figure 1),
in which the accident probability was correlated with the
number of doors.

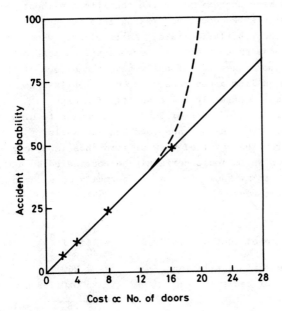

Figure 1

The latter seemed to have been recommended on the basis of
movement distances drawn on plans with little account taken of
any of the other parameters. The graph of accident probability
against number of doors would be the straight line shown in

Figure 1. In fact, the nearer the doors are to each other
the more they interact, thus increasing the accident
probability. The true accident probability is depicted much
more accurately by the dotted line in Figure 1. To examine a
particular case we considered the effect on a volume of air
trapped between doors positioned on either side of a service
hatch communicating with chemical stores. We assumed the
occurrence of an accident which is fairly common in chemical
laboratories, i.e. in this case dropping of a winchester bottle,
which can smash even if it is in an approved carrier.
Calculating the amount of vapour from, in this case, benzene,
which we took as an example typical of both a toxic and a
flammable material, we made the calculations shown.
Calculations were made approximating the conditions without
smoke doors and with smoke doors and taking into account the
modifications of the ventilation rate. As is clearly shown in
Table 1, although the smoke doors did serve their purpose of
limiting fire or smoke they additionally have the effect of
converting this area from a non-explosive condition into an
explosive condition. Certainly the quantity of energy
liberated, which is of the order of 30kJ, is adequate to blow
out windows, doors, and possibly non-load-bearing walls.
Obviously, even from the point of view of fire this is a very
serious interaction which would certainly be considered by a
chemist, but which to my knowledge is not normally considered
by the enforcing agencies.

Table 1

Total volume of section 9.8.20	$= 1440 \text{ ft}^3$
	$= 40,800 \text{ l}$
With 5 minutes' ventilation	$= 54,260 \text{ l}$
Winchester of benzene	$= 2500 \times 0.789 \text{ g}$
	$= \dfrac{2500 \times 0.789}{78} \text{ mole}$
Vapour at 20 $^\circ$C, 760 mm	$= \dfrac{2500 \times 0.789}{78} \, 24 \text{ l}$
V/V% benzene in atmosphere – no vent	$= \dfrac{607,100}{40,800}$
	$= 149 \text{ V/V%}$
V/V% benzene in atmosphere – with vent	$= \dfrac{607,100}{54,260}$
	$= 112 \text{ V/V%}$

Lower explosive limit for benzene in air at 20 $^\circ$C is 140 V/V%

It seemed profitable to examine the proposition that in places such as chemical laboratories handling toxic materials, where in emergency situations it was neither practicable nor necessarily safe to discontinue the ventilation, the position of fire stops was influenced much more by the direction and rate of ventilation than by travel distance. In order to examine this proposition we borrowed a building for a weekend - the Department of Physics, University of Bristol - a building typical of between-the-wars construction and also of the type commonly subjected to modification at the present time. The layout is as shown in Figure 2.

Hot smoke was discharged in various parts of the building with the ventilation switched on and off and in the absence of smoke doors. The procedure was repeated with simulated smoke doors in position. The behaviour of the smoke in the building was perfectly straightforward and could have been predicted. The very large stairwell acted as a thermodynamic sink and was of sufficiently large volume, which, combined with good mixing at the base, absorbed without any major inconveniences the smoke and hot gases produced. At no time in the absence of smoke doors was there any difficulty in escaping from the building.

However, once the doors were in position smoke was, in some cases, forced up hitherto unused passages (i.e. lift shafts) to block floors which had been untouched previously. In some cases, escape from the corridors on which smoke was liberated became impossible after a minute and a half. Obviously, this is a very worrying experimental result, although it is in no way suggested that smoke doors are not useful; results in other buildings in the University indicated the beneficial effects of correctly placed smoke doors. It did, however, stress that the position of safety devices in buildings should be considered in relation not just to travel distances but to ventilation rates and directions, work population, and population densities. In other words, the problem should be considered as a whole in relation to what is happening in each individual building. Only then can the optimum safety conditions be ascertained.

CORRIDORS, PHYSICS OLD BUILDING.

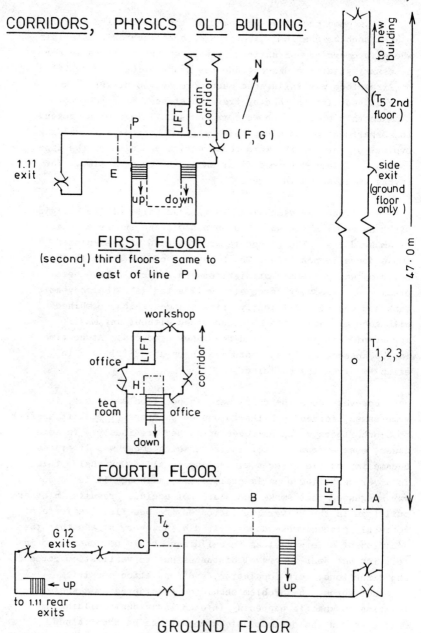

Figure 2   Corridors, Physics Old Building

As a final example and as a piece of black humour, I would like to refer to a door. This door separates a working laboratory from instrument workshops and offices on a corridor. Despite initial design agreements with the architects somebody, possibly the Building Regulations Inspector etc., had achieved the requirement of a two-hour fire door at a cost in excess of £1000. This door was already too heavy for the hinges on which it hung. Since one could not contemplate a door to a laboratory of this type without a vision panel the response was to cut a vision panel in it, thus reducing the efficiency of the door to half an hour. This door could have been bought and installed for less than £300 from the start and would actually have hung correctly. Furthermore, because the door had to be self-closing, in order to allow the fume cupboards and ventilation systems within the lab actually to obtain a supply of air, ventilation panels were then cut into the door at further high cost. In order apparently to satisfy some mythical fire requirements these were made with an intumescent filling so that they would expand and produce a barrier in the case of fire. The main problem in such a laboratory is, however, either from fast fire, i.e. explosion, or release of toxic gases, probably cold; this device has absolutely no effect and renders the whole design an expensive piece of uselessness. So much for reality in design.

The foregoing examples illustrate the problems which can occur when one ignores the interactions which occur in real working environments. Legislation aimed at controlling one hazard must be sufficiently flexible in its approach to allow such interactions with other problems to be taken into account. Solutions to hazard problems should be considered not only in their ability to solve the matter at hand but also in the light of their possible enhancement of hazards in other areas. Thus, the aim of any legislation should be the cost-beneficial control of the general safety standards. Solutions that, although controlling one particular problem, may actually increase the overall hazard situation cannot be acceptable. It would seem that in order to obtain such legislation a much greater input of working expertise, and in particular scientific knowledge, should be made before such legislation

is passed. Legislation by regulation should only be made
when control by a Code of Practice is completely out of the
question.

It is gratifying to recognise that in the past two or
three years the Royal Society of Chemistry through its
committees has taken a much stronger line in the area of
safety legislation and is continuing to do so. It is my
belief that professional bodies such as the Royal Society of
Chemistry should continue to make a powerful contribution aimed
at affecting such legislation before it is made.

As a final warning I would draw your attention to the
forthcoming Ionising Radiations Regulations and the draft
document that has already been circulated and which is causing
some concern amongst those who have studied it.

Finally, the considerations illustrated by the examples
discussed above lead to the formulation of a general rule:

> Good safety legislation, whether by regulation
> or Code of Practice or other means, should
> reduce the risk not only in that area at which
> it is principally directed, but should as a
> consequence, lead to a general reduction in
> risk in all fields.

No Discussion

The Protection of Workers Exposed to Chemicals: the European Community Approach

By Dr W. Hunter*, A. Berlin, M. Th. Van der Venne, and A. E. Bennett
HEALTH AND SAFETY DIRECTORATE DG V.E.2, COMMISSION OF THE EUROPEAN
COMMUNITIES, LUXEMBOURG

Background to the Community's Actions

The three treaties creating the European Communities
(Coal and Steel, Euratom, and the Economic Community), which
were signed in the 1950s, all proclaim the need to achieve
better health and wellbeing for the citizens of the Member
States, and particular attention is devoted to the health
and safety of workers.

In 1974 the Council of Ministers decided to strengthen
the activities in health and safety at work by establishing
a committee entitled the Advisory Committee on Safety, Hygiene
and Health Protection at Work, with a tripartite composition:
two members from government, two from employers, and two
from workers, making a total of six members from each Member
State.

In 1978 the Council took a major step forward in furthering
the protection of workers by adopting the first action
programme[1] of the European Communities on safety and health
at work to run until the end of 1982. This programme set out
the basis for a common policy aimed at increasing the
protection of more than 100 million workers' health and
safety throughout the ten Member States. It contained fourteen
actions, of which six were dedicated to the protection of
workers from the harmful effects of chemicals. The actions
of the first programme have led to several important initiatives
being taken at Community level.

In view of its success, the Commission has now proposed

a second programme[2], which builds on and extends the actions
provided for in the Community's first programme.

The new programme, which is expected to be approved at
a Council of Ministers' session in 1983, covers the areas
of protection against dangerous agents, ergonomic measures,
protection against accidents and dangerous situations,
organisation, training and information statistics, research,
and co-operation. The programme provides for a continuation
of the work already in progress and for certain new actions
to 'reflect the changed needs and concerns of today's society'.
The programme is intended to run for six years - from 1983
until the end of 1988.

Protection against Dangerous Agents - The first actions in
the new programme are concerned with the protection of workers
against dangerous agents at work.

It provides for:

- Common methodologies to be established for the
  assessment of the health risks of physical, chemical,
  and biological agents present at the workplace;

- A 'standard approach' to be developed for establishing
  exposure limits for toxic substances, and recommendations
  to be made for harmonising exposure limits for a certain
  number of substances, taking into account existing
  exposure limits;

- Standard methods of measuring and evaluating concentrations in the air and biological indicators at the
  workplace;

- 'Preventive and protective' actions to be developed
  for agents 'recognised' as being carcinogenic and
  other dangerous substances and processes which may
  produce serious health effects;

- Continuation of work on vibrations and other non-
  ionising radiations.

Ergonomic Measures, Protection against Accidents and
Dangerous Situations - The second set of actions in the new
programme is concerned with basic ergonomics, workplace safety,
and specific accident prevention measures - issues not dealt
with in any detail in the first action programme. The new
programme provides for:

- Proposals for safety measures, particularly for certain
  high-risk activities, and for the prevention of
  accidents involving falls and manual lifting and
  handling;

- Examination of the major accident hazards of certain
  industrial activities covered by Council Directive
  82/501/EEC of 24 June 1982;[8]

- Working out ergonomic measures and accident prevention
  principles relating to the design of equipment;

- Monitoring and improving the effectiveness of safety
  and health protection by organising exchanges of
  experiences with a view to establishing more clearly
  the principles and methods of organisation and training
  followed for and by those responsible in the fields of
  safety, health, and hygiene at work.

Organisation - The third set of actions concerns certain
organisational aspects of workplace health and safety. The
programme provides for:

- Recommendations on the 'organisational and advisory role'
  of the departments in charge of health and safety in
  small and medium-sized industries by defining in
  particular the role of specialists in occupational
  health, hygiene, and safety.

- Drawing-up the principles and criteria for monitoring
  workers whose health and safety is likely to be 'at
  high risk' such as certain maintenance and certain
  repair workers, migrant workers, and others in sub-
  contracting undertakings.

Training and Information - The fourth set of actions in the
new programme concerns general collection and dissemination
of information and safety training. The programme provides
for:

- The preparation of information notices and manuals
  on the handling of certain dangerous substances,
  particularly those covered by Community Directives;

  The drawing-up of proposals on the establishment of
  systems and codes for the identification of dangerous
  substances at the workplace;

- The drawing-up of programmes and training programmes.

Statistics, Research, and Co-operation - The final actions
in the new programme deal with improving co-operation and
statistical and research data in the health and safety field.
The programme provides for:

- The development of comparable data on mortality
  connected with work and the collection of other data
  on the frequency, gravity, and causes of accidents
  at work and occupational diseases;

- An inventory to be compiled of the 'cancer registers'
  currently existing at local, regional, and national
  levels, so as to assess the comparability of data
  contained in them and to ensure better co-ordination
  at Community level;

- The identification and co-ordination of topics for
  applied research in health and safety at work which

may be the subject of future Community action;

- Continued co-operation with international organisations such as the World Health Organisation and the International Labour Office, and with national organisations and institutes outside the Community;

- Continued co-operation with other actions by the Community and by Member States.

Protection against Dangerous Agents

'Framework' Directive - An essential step in the implementation of the first programme was the adoption by Council on 27 November 1980[3] of a Directive on the Protection of Workers against the risks connected with exposure to chemical, biological and physical agents at work. This is a broad 'framework' Directive which should result in all Member States following a similar path in the future. This Directive sets out two objectives:

- The elimination or limitation of exposure to chemical, physical, and biological agents and the prevention of risks to workers' health and safety;

- The protection of workers who are likely to be exposed to these agents.

The Directive requires the Member States to take short- and longer-term measures; it also foresees the adoption by Council of individual Directives laying down limit value(s) and other specific requirements for named Agents.

The short-term measures require that within three years appropriate information be provided to workers and/or their representatives concerning the health risks due to asbestos, cadmium, lead, and mercury, and that within four years appropriate surveillance be set up of the health of workers exposed to asbestos and lead.

The longer-term measures apply when a Member State adopts
provisions concerning an agent. In order that the exposure
of workers to agents is avoided or kept at as low a level as
is reasonably practicable, Member States should comply with
a set of requirements, but in doing so they have to determine
whether and to what extent each of these requirements is
applicable to the agent concerned. Some of the most important
of these requirements are:

- Limitation of use at the place of work;

- Limitation of the number of workers exposed;

- Prevention by engineering control;

- Establishment of limit values and of sampling and
  measuring procedures, and methods for evaluating results;

- Protection measures involving the application of suitable
  working procedures and methods;

- Collective and individual protection measures, where
  exposure to agents cannot be avoided by the other means,
  as well as hygiene measures;

- Emergency procedures for abnormal exposures;

- Information for workers;

- Surveillance of the workers' health.

In addition further, more specific requirements are laid down
for a list of eleven substances. These are:

- Provision of medical surveillance of workers by a doctor
  before exposure and thereafter at regular intervals;

- Access by workers and/or their representatives at the
  workplace to the results of ambient and biological
  (collective) exposure measurements;

- Access by each worker concerned to the results of
his own biological tests indicating exposure;

- Information of workers and/or their representatives
at the workplace of cases where the limit values are
exceeded, of the causes thereof and of the measures
taken to rectify the situation;

- Access by workers and/or their representatives to
appropriate information to improve their knowledge
of the dangers to which they are exposed.

The Directive also requires the Member States to consult the
social partners when the above requirements are being established.

Safety Signs - This Directive was adopted in 1977 even before
the first programme of action on safety and health was adopted.
It relates to the provisions of safety signs at places of work.[4]
Certain general principles have been laid down in this Directive,
namely:

1. The objective of the system of safety signs is to
   draw attention rapidly and unambiguously to objects
   and situations capable of causing specific hazards.

2. Under no circumstances is the system of safety signs
   a substitute for the requisite protective measures.

3. The system of safety signs may be used only to give
   information related to safety.

4. The effectiveness of the system of safety signs is
   dependent in particular on the provision of full and
   constantly repeated information to all persons likely
   to benefit therefrom.

The Directive lays down the:

- Safety colours and contrasting colours of safety signs;

- Geometrical form and meaning of safety signs;

- Combinations of shapes and colours and their meanings
  for signs;

- Design of safety signs;

- Yellow/black danger identification.

The special system of safety signs is divided into four
groups:

   (i)   Prohibition signs
   (ii) Warning signs
   (iii) Mandatory signs
   (iv) Emergency signs.

Examples of these signs are given in Annex 1.

<u>Vinyl Chloride Monomer</u> - The Commission acted rapidly following
the publication of the studies regarding vinyl chloride monomers
and angiosarcomas. In 1975 a first scientific and technical
meeting was convened in Brussels during which technical limits
were considered.

During the preparation of the Community proposal for a
Directive further evidence accumulated on the carcinogenic
effect of vinyl chloride monomer on man while at the same time
considerable efforts were being made at the technical level to
reduce worker exposure.

The resulting Directive, which was adopted in June 1978,[5]
is the first to deal with the control of worker exposure to a
chemical carcinogen. It contains the following main provisions:

- Establishment of atmospheric limit values and the
  fixing of provisions for monitoring;

- Technical preventive measures;

- Information of workers;

- Keeping a register of exposed workers;

- Guidelines regarding medical surveillance.

The Directive covers all workers employed in works in which
vinyl chloride monomer is produced, reclaimed, stored,
discharged into containers, transported, or used in any way
whatsoever, or in which vinyl chloride monomer is converted
into vinyl chloride polymers.

The technical long-term (one year) limit value established
is of 3 ppm., with equivalent limit values for shorter periods
of time (see Table 1). In general both continuous and
discontinuous methods may be used for monitoring the vinyl
chloride monomer concentration in a working area, so long as
the measurement systems have a sensitivity of at least 1 p.p.m.
In the case of vinyl chloride polymerisation plants only
continuous monitoring is acceptable.

Table 1 - Limit values for vinyl chloride in parts per million
in relation to reference periods

Reference period	Limit value (p.p.m.)
(1 year)	(3)
1 month	5
1 week	6
8 hours	7
1 hour	8

Personal protection measures have to be taken in case of
abnormal increases in vinyl chloride monomer concentration
levels, and thresholds are also laid down in the Directive.
These protection measures have to be taken also for certain
operations such as the cleaning of autoclaves.

A register of exposed workers has to be kept for at least
30 years and has to contain particulars of the type and duration

of work and exposure to which they have been subjected.

Medical examinations are required but the frequency
and type of medical examinations have to be determined by
each Member State. However, guidelines for the medical
examinations are provided in the Directive as are the basic
requirements for examinations before exposure.

Individual Directives - In the Framework Directive,[3] eleven
agents (Table 2) are named for early action by means of
individual Directives laying down limit values and other
specific requirements. The Commission has already made
proposals on two of them, namely lead[6] and asbestos[7]. In
July 1982 the Council adopted the Directive on lead, this
being the first of the individual Directives.

Table 2 - List of agents named in the 'Framework Directive'

        Acrylonitrile
        Asbestos
        Arsenic and compounds
        Benzene
        Cadmium and compounds
        Mercury and compounds
        Nickel and compounds
        Lead and compounds
        Chlorinated hydrocarbon compounds: chloroform;
                                            paradichlorobenzene;
                                            carbon tetrachloride

Lead

This Directive requires that all work presenting a risk
of absorption of lead be assessed to determine the nature and
degree of exposure to lead. The lead in air and the blood
lead concentrations in relation to the actions to be taken
as regards these concentrations are given in Table 3.
Regular lead in air monitoring must be representative of
workers' exposure and a number of technical details regarding
the sampling procedure are specified. Biological monitoring,
essentially blood lead measurement, is to be performed on all
workers every six months and clinical surveillance shall be
carried out at least once a year.

Table 3 - <u>Lead in air and blood lead concentrations in</u>
<u>relation to actions to be taken</u>

<u>Lead in air</u> <u>concentrations</u> $\mu g\ m^{-3}$ <u>(40 hours per week</u> <u>time-weighted average</u>)	<u>Blood lead</u> <u>concentrations</u> $\mu gPb/100ml$ <u>blood</u>	<u>Actions to be taken</u>
40	40	Minimise lead absorption of workers Provide information to workers
	40-50	Regular biological monitoring considered appropriate
75	50	All the protective measures of the Directive apply including: - Lead in air monitoring - Medical surveillance
150	70	Limit values requiring action to reduce exposure
	70-80	Limit value may be acceptable if other biological indication below certain limits (<u>e.g.</u> ALAU 20 mg/g creatinine).

Several hygiene measures are laid down, including the
limitation of risk of absorption of lead through smoking,
drinking, and eating; the provision of special work clothes;
and of adequate washing facilities. Provisions are also
included regarding the need to inform workers of the dangers
to health due to lead exposure, of the need for appropriate
protective measures, and of the monitoring results (individual
lead in air values and biological group values), as well as
the need to consult workers on lead in air monitoring procedures
and on the measures to be taken when the lead in air limit
values are exceeded. Access to information must also be given

to the doctor or authority responsible for the medical
surveillance of workers in order to evaluate the exposure
of workers to lead.

The 'lead' Directive allows Member States to introduce
more stringent protective measures in their national
legislation either for all workers or for certain categories
of workers. This enables special account to be taken of the
potential risk to the embryo and to the development of the
foetus which may result from a high body burden of lead.

Thus this Directive sets out a comprehensive set of
measures aimed at improving the protection of workers
against the risks to health exposure in all Member States.

Asbestos - The importance of the health risk due to asbestos
must not be underestimated. In addition to asbestosis,
asbestos is associated with cancer of the lung, mesotheliomas,
and cancers of the gastro-intestinal system. Within the
European Community the number of deaths from mesothelioma
probably exceeds 700 per annum.

The proposal for an individual Directive on asbestos[7]
is currently still under discussion at the Council, but the
progress that has been made suggests that it will be adopted
in 1983. Both sides of the industry, as represented by the
Asbestos International Association and by the European Trade
Union Conference, have expressed their desire to see this
proposal adopted as quickly as possible.

In common with the lead Directive,[6] the asbestos
proposal[7] puts forward limit values and other specific
requirements. The proposed limit values are 0.2 fibres/ml
(2×10^4 fibres/m^3) for crocidolite and 1 fibre/ml (1×10^5
fibres/m^3) for non-crocidolite fibres measured or
calculated over a reference period of 8 hours. A fibre
is defined as having a length greater than 5 μg,
with a length to breadth ratio greater than three.

A sampling and analysis strategy is put forward,

including the use of the membrane filter method and
optical microscopy.

In addition proposals are made for:

- The register of working sites or plants;
- The prohibition of spraying of asbestos;
- The precautions to be taken in working with
  asbestos-containing materials;
- The conditions under which asbestos can be
  handled and transported;
- Demolition and dismantling work, where this
  involves asbestos or asbestos-containing materials.

Proposals are also made regarding:

1. The information of workers so that they are:

   - Properly instructed on correct handling procedures
      and informed of potential risks;
   - Supplied with appropriate protective clothing and
      safety devices.

2. The cleaning and maintenance of protective clothing
   and equipment.

3. The requirements for regular medical examinations and
   the keeping of adequate medical records.

Prevention of Major Accidents

In the past decade a significant number of major chemical
accidents have occurred both within the European Community and
in other countries.

Accidents such as those of Flixborough, Seveso, and Manfedonia
have prompted not only the Member States but also the Community to
act. Thus in June 1982, Council adopted a Directive on major
accidents hazards of certain industrial activities.[8] This aimed
at preventing such accidents and reducing any consequences if

they occur. Owing to its importance for the workers, the general population, and the environment, the Directive is based on both the Action Programme on Safety and Health at Work[1] and on the Environmental Action Programme.[9]

A major accident is defined as:

A major emission, fire, or explosion involving one or more dangerous substances, resulting from the uncontrolled development of an industrial activity, which could constitute a serious risk, immediate or delayed, for workers, the neighbouring population, and the environment.

The Directive requires that for industrial activities which involve dangerous substances:

- A safety report be drawn up;
- Workers be informed, equipped, and trained;
- Safety drills be organised;
- The neighbouring population be informed and an emergency plan established.

In addition when particularly toxic, persistent, explosive, or flammable substances are present in quantities exceeding certain limits then notification with a more substantial dossier is compulsory. Table 4 gives some examples.

Table 4 - Examples of substances requiring notification

Substance	Quantity
2-Acetylaminofluorene	1 kg
Benzidine	1 kg
Beryllium oxide	1 kg
Arsenic (solid compounds)	500 kg
Phosgene	20 t
Chlorine	100 t
Carbon disulphide	200 t
Ammonia	1000 t
Hydrogen	20 t

          Ammonium nitrate   5000 t
          Liquid oxygen   10000 t

The Directive requires that national authorities inspect
periodically any notified industrial activities and that the
manufacturer informs the national authorities, according to a
given format, of any accidents.

In addition Member States are immediately to inform the
Commission of any major accidents occurring on their territory,
and the Commission will set up a data bank on such accidents
for the mutual information of Member States.

Dangerous Substances - Testing

In 1967 the Commission of the European Community submitted
to the Council of Ministers a proposal, based on the work
done at international level, for a Directive concerning the
approximation of the laws, regulations, and administrative
provisions relating to the classification, packaging, and
labelling of dangerous substances.

Since 1967, the concept of dangerous substances has
significantly evolved. Thus, the Sixth Amendment (79/831/EEC),
with its pre-marketing notification requirement of new chemicals,
also contains provisions relating to classification and labelling
of both new and existing substances (including imported chemicals)
placed on the Community market.[10]

The EEC labelling requirements are intended to provide a
clear primary means by which all persons (workers as well as
the public at large) handling or using substances are given
essential information about the inherent dangers of the
labelled substances. The EEC label is intended to give
information on two types of danger, health dangers and physical
dangers.

Up to now approximately 1000 substances have been examined,
classified, and listed. However this total of 1000 substances
is very small when compared with the number of existing
substances, which total about 35,000.

This lack of information is the main reason for the fundamental modification in approach which has taken place with the Sixth Amendment.

For <u>existing substances</u> classification and labelling must take place in so far as the manufacturer or any person who places these substances on the market may reasonably be expected to be aware of their dangerous properties. The data required for classification and labelling of these substances may have to be derived from different sources, for example previous test data, information required in relation to international rules of dangerous goods, and information obtained from the literature or derived from practical experience.

For <u>new substances</u> classification and labelling is mandatory and will be based on the data submitted to the Competent Authorities in the notification dossier. This dossier will include:

- A technical dossier supplying the information necessary to evaluate the risks which the new substance may entail for man and the environment. It should contain at least the information and results of the studies referred to in the so called 'base set' which concerns physicochemical, toxicological, and ecotoxicological tests; Table 5 shows the toxicity tests as foreseen in the base set;

- A declaration concerning the unfavourable effects of the substance in terms of the various uses envisaged;

- The proposed classification and labelling;

- The proposals for any recommended precautions relating to the use of the substance.

The results of all these tests will thus allow not only a more appropriate classification of the substance, but will also serve as one of the sources of information for setting control requirements as well as indicating the appropriate hygiene and medical surveillance measures that need to be taken for workers exposed to these substances.

Table 5 - <u>Base set of toxicity tests</u>

1. <u>Acute toxicity</u>

    1.1 LD50 oral, inhalation, cutaneous.
         Usually two routes of administration.
         Rats male and female.
         14 days observation.

    1.2 Skin irritation
         Albino rabbit

    1.3 Eye irritation
         Rabbit

    1.4 Skin sensitisation
         Guinea-pig

2. <u>Sub-acute toxicity</u>

    28 day administration. Usually oral,
    preferably rat

3. <u>Mutagenicity</u>

    Series of two tests.
    Bacteriological with and without metabolic activation
    Non-bacteriological

<u>Health Risk Evaluations and Toxicological Training</u>

The provisions of the Sixth Amendment Directive[10] and the need
to carry out health risk evaluations spelled out in the Action
Programme on Safety and Health at Work[1] have led the Commission
into establishing standard protocols for a series of toxicological
tests and to develop Good Laboratory Practice Guidelines. This
should allow the mutual recognition of toxicological data, with
the consequence of reducing, if not eliminating completely,
unnecessary tests. The Commission is also actively examining
currently available alternative methods to animal testing

together with research requirements.

Recently a survey was completed on existing university
programmes in toxicology and other fields requiring some
knowledge of toxicology. Based on the results of this survey
it is intended to develop appropriate recommendations on the
teaching of toxicology and to suggest minimum curricula for
such training.

Other initiatives
There are three initiatives which deserve mention:

1. There are certain agents that have long been
   identified as being particularly dangerous
   e.g. 2-naphthylamine. The Commission is
   currently drawing up a proposal aimed at
   protecting workers from the risks related to
   exposure to certain agents or work activities
   by means of prohibition. The proposal is at
   an advanced stage of preparation and at present
   covers 2-naphthylamine, 4-aminobiphenyl and
   4-nitrobiphenyl. The Commission anticipates
   that this proposal will be sent to the Council
   in the course of 1983.

2. The Commission is currently evaluating the data
   available on the agents in Table 2 with a view
   to preparing other individual Directives. It
   is anticipated that at least one of these proposals
   will be sent to the Council in 1984.

3. As regards the protection of workers from physical
   agents, the Commission has already made two proposals
   for Council Directives, the first on microwaves[11]
   and the second on noise.[12]

References

1. 'Council Resolution of 28 June 1978 regarding an
    Action Programme on Safety and Health at Work',
    Official Journal of the European Communities,
    11.7.1978, C165, 1.

2. 'Proposal for a Council resolution on a Second
    Programme of Action of the European Communities
    on Safety and Health at Work', Official Journal
    of the European Communities, 25.11.1982, C308, 11.

3. 'Council Directive (EEC/1107/80) of 27 November
    1980 for the Protection of Workers from the Risks
    of Exposure to Chemical, Physical and Biological
    Agents at Work', Official Journal of the European
    Communities, 3.12. 1980, L327, 8.

4. 'Council Directive (77/576/EEC) of 25 July 1977 on
    the approximation of the laws regulations and
    administrative provisions of safety signs at places
    of work', Official Journal of the European Communities,
    7.9.1977, L299, 12.

5. 'Council Directive (78/610/EEC) of 29 June 1978 on
    the approximation of the laws, regulations and
    administrative provisions of the Member States on
    the protection of health of workers exposed to vinyl
    chloride monomer', Official Journal of the European
    Communities, 22.7.1978, L197,12, and 7.9.1977, 12.

6. ' Council Directive (EEC/605/82) of 28 July 1982 on
    the Protection of Workers from Harmful Exposure to
    Metallic Lead and its ionic compounds at work',
    Official Journal of the European Communities, 23.8.1982,
    L247, 12.

7. 'Proposal for a Second Council Directive on the Protection
    of Workers from the Risks related to Exposure to Agents
    at Work: Asbestos', Official Journal of the European
    Communities, 9.10.1980, C262, 7.

8. 'Council Directive (82/501/EEC) of 24 June 1982
    on the major accident hazards of certain industrial
    activities', Official Journal of the European
    Communities, 5.8.1982, L230, 1.

9. 'Second Action Programme of the European Communities
    on the Environment of 17 May 1977', Official Journal
    of the European Communities, 13.6.1977, C139, 1.

10. 'Council Directive (79/831/EEC) of 18 September 1979
    amending for the sixth time Directive 67/548/EEC on
    the approximation of the laws, regulations and
    administrative provisions relating to the classification,
    packaging and labelling of dangerous substances',
    Official Journal of the European Communities, 15.10.1979,
    L259, 10.

11. 'Proposal for a Council Directive laying down basic
    standards for the health protection of workers and
    the general public against the dangers of microwave
    radiation', Official Journal of the European Communities,
    16.9.1980, C249, 6.

12. 'Proposal for a Council Directive on the protection
    of workers from the risks related to exposure to
    chemical, physical and biological agents at work:
    noise', Official Journal of the European Communities, 5.11.12,
    C289, 1.

Annex 1

SÆRLIG SIKKERHEDSSKILTNING — BESONDERE SICHERHEITSKENNZEICHNUNG — SPECIAL SYSTEM OF SAFETY SIGNS — SIGNALISATION PARTICULIÈRE DE SÉCURITÉ — SEGNALETICA PARTICOLARE DI SICUREZZA — BIJZONDERE VEILIGHEIDSSIGNALERING

1. Forbudstavler — Verbotszeichen — Prohibition signs — Signaux d'interdiction — Segnali di divieto — Verbodssignalen

a)

Rygning forbudt
Rauchen verboten
No smoking
Défense de fumer
Vietato fumare
Verboden te roken

b)

Rygning og åben ild forbudt
Feuer, offenes Licht und Rauchen verboten
Smoking and naked flames forbidden
Flamme nue interdite et défense de fumer
Vietato fumare o usare fiamme libere
Vuur, open vlam en roken verboden

c)

Ingen adgang for fodgængere
Für Fußgänger verboten
Pedestrians forbidden
Interdit aux piétons
Vietato ai pedoni
Verboden voor voetgangers

d)

Sluk ikke med vand
Verbot, mit Wasser zu löschen
Do not extinguish with water
Défense d'éteindre avec de l'eau
Divieto di spegnere con acqua
Verboden met water te blussen

e)

Ikke drikkevand
Kein Trinkwasser
Not drinkable
Eau non potable
Acqua non potabile
Geen drinkwater

2. Advarselstavler — Warnzeichen — Warning signs — Signaux d'avertissement — Segnali di avvertimento — Waarschuwingssignalen

a)

Brandfarlige stoffer
Warnung vor feuergefährlichen
Stoffen
Flammable matter
Matières inflammables
Materiale infiammabile
Ontvlambare stoffen

b)

Eksplosionsfarlige stoffer
Warnung vor explosionsgefährlichen
Stoffen
Explosive matter
Matières explosives
Materiale esplosivo
Explosieve stoffen

c)

Giftige stoffer
Warnung vor giftigen Stoffen
Toxic matter
Matières toxiques
Sostanze velenose
Giftige stoffen

d)

Ætsende stoffer
Warnung vor ätzenden Stoffen
Corrosive matter
Matières corrosives
Sostanze corrosive
Bijtende stoffen

e)

Ioniserende stråling
Radioaktivitet/Røntgenstråling
Warnung vor radioaktiven Stoffen oder
ionsisierenden Strahlen
Radioactive matter
Matières radioactives
Radiazioni pericolose
Radioactieve stoffen

f)

Kran i arbejde
Warnung vor schwebender Last
Beware, overhead load
Charges suspendues
Attenzione ai carichi sospesi
Hangende lasten

g)

Pas på kørende transport
Warnung vor Flurförderzeugen
Beware, industrial trucks
Chariots de manutention
Carrelli di movimentazione
Transportvoertuigen

h)

Farlig elektrisk spænding
Warnung vor gefährlicher elektrischer
Spannung
Danger: electricity
Danger électrique
Tensione elettrica pericolosa
Gevaar voor elektrische spanning

i)

Giv agt
Warnung vor einer Gefahrenstelle
General danger
Danger général
Pericolo generico
Gevaar

3. Påbudstavler — Gebotszeichen — Mandatory signs — Signaux d'obligation — Segnali di prescrizione — Gebods-signalen

a)

Øjenværn påbudt
Augenschutz tragen
Eye protection must be worn
Protection obligatoire de la vue
Protezione degli occhi
Oogbescherming verplicht

b)

Hovedværn påbudt
Schutzhelm tragen
Safety helmet must be worn
Protection obligatoire de la tête
Casco di protezione
Veiligheidshelm verplicht

c)

Høreværn påbudt
Gehörschutz tragen
Ear protection must be worn
Protection obligatoire de l'ouïe
Protezione dell'udito
Gehoorbescherming verplicht

d)

Åndedrætsværn påbudt
Atemschutz tragen
Respiratory equipment must be used
Protection obligatoire des voies respira-
toires
Protezione vie respiratorie
Adembescherming verplicht

e)

Fodværn påbudt
Schutzschuhe tragen
Safety boots must be worn
Protection obligatoire des pieds
Calzature di sicurezza
Veiligheidsschoenen verplicht

f)

Beskyttelseshandsker påbudt
Schutzhandschuhe tragen
Safety gloves must be worn
Protection obligatoire des mains
Guanti di protezione
Veiligheidshandschoenen verplicht

4. Redningstavler — Rettungszeichen — Emergency signs — Signaux de sauvetage — Segnali di salvataggio — Reddings-
signalen

a)

Førstehjælp
Hinweis auf „Erste Hilfe"
First aid post
Poste premiers secours
Pronto soccorso
Eerste hulp-post

b)

c)

eller/oder/or/ou/o/of

d)

Retningsangivelse til nødudgang
Fluchtweg (Richtungsangabe für Flucht-
weg)
Emergency exit to the left
Issue de secours vers la gauche
Uscita d'emergenza a sinistra
Nooduitgang naar links

e)

Nødudgang
(anbringes over udgangen)
Fluchtweg
(über dem Fluchtausgang anzubringen)
Emergency exit
(to be placed above the exit)
Sortie de secours
(à placer au-dessus de la sortie)
Uscita d'emergenza
(da collocare sopra l'uscita)
Nooduitgang
(te plaatsen boven de uitgang)

Discussion

Q1.   Can you tell us whether funding is available
         from the community for research projects on
         health and safety?

Hunter - Yes funding is available for either study projects
          or research proposals. In the case of the studies,
          in particular RSC's own morbidity and mortality
          studies, Barry Henman came over to talk to us some
          years ago about whether they could be set up.
          As a result of this RSC initiative a lot of funding
          has been provided over the past six years.

          This type of funding activity is not heavily
          advertised - it is one of our quieter activities.
          The budget for the whole of the programme of action
          on safety and health at work is about two million
          units of account (about £80 million). We have a
          certain amount of discretion as to how we use it.

          As regards specific research proposals there are
          two types:-

          (i)   Funded research. Here about half of the
                cost of a particular proposal can be funded.

          (ii) Concerted research. Here various countries
                undertake to carry out parts of a research
                program so that the results can be dovetailed
                together. Money is available for coordinating
                the research, but not funding it.

          As far as research is concerned calls for tenders
          are published in the Official Journal of the
          European Communities.

Q2.   In the various Community Directives relating to
         health and safety are there provisions for the

education, training and ethical conduct of
individuals?

Hunter -   This is very complicated.

It is a feature of the Community that directives
sometimes have repercussions. For example, a
directive relating to the free movement of medical
doctors went through a few years ago. This
directive had 'created' the speciality of
occupational medicine. This directive is now
being revised and requirements for the training
of occupational health physicians will be laid
down. In other directives on free movement of
labour the question of various other groups (eg.
chemists, pharmacists) has come up. However the
directive concerning doctors, including occupational
health physicians, is at a much more advanced state
than those of other professional groups.

The sixth amendment of the dangerous substances
directive sets up a notification scheme for new
chemical substances for which competent toxicolo-
gists are required. We are surveying toxicology
and training in toxicology in order to determine
the current situation.

Recommendations Arising from the Symposium

I   There is an immediate and continuing need for chemical societies to

- make the general public more aware that professional chemists are actively engaged in making chemistry safer

- counter and discourage adverse media comments regarding health and safety in chemistry.

II   There is a longer-term need for chemical societies to participate in the development of

- international harmonisation of voluntary guidelines on health and safety

- international standardisation of definitions of accidents and of accident statistics together with a coordinated approach to obtaining detailed statistics on accidents and dangerous occurrences

- national standardisation (by governments if on a mandatory basis or by independent bodies if on an advisory basis) of evaluation of the cost effectiveness of control measures regarding health and safety

- common procedures and minimum standards
  (possibly through special protocol
  committees) for health and safety in the
  workplace

- common procedures for the identification of,
  and reporting on, the incidence of
  ill-health from work-related disease

- training programmes in health and safety
  (possibly carried out by professional bodies)
  to be made available to small businesses.

III   There is a longer-term need for

- an internationally accepted definition of a
  carcinogen to be established

- morbidity and mortality studies to be
  conducted among professional workers other
  than chemists and among technical and manual
  workers; and for trades unions to conduct
  such studies among their members

- reduced premiums to be charged by insurance
  companies to organisations with good health
  and safety records

- greater collaboration on health and safety
  among professional bodies in the fields of
  chemistry and toxicology

- greater awareness of toxicology to be
  included in the training of chemists.